FROM FOREST TO SEA

Eric Rolls was born at Grenfell in western New South Wales in 1923 and spent his childhood on a farm outside Narrabri. During World War II he joined the 24th Light Horse, and then went to Papua New Guinea with the New Guinea Air Warning Wireless Company to report movements of Japanese troops and aeroplanes. While in the army he sent the poem "Death Song of a Mad Bush Shepherd" to Douglas Stewart who published it in the *Bulletin*. The poem was broadcast by the ABC and later by the BBC.

After the war, Eric Rolls farmed at Boggabri and then at Baradine and continued to write. He now lives in Sydney and writes full-time. His award-winning publications include *Sheaf Tosser, They All Ran Wild, Running Wild, The River, A Million Wild Acres, Celebration of the Senses, Doorways: A Year of the Cumberdeen Diaries, Selected Poems* and *Sojourners: The story of China's centuries-old relationship with Australia.* He is currently working on *Citizens*, the sequel to *Sojourners*, as well as researching *The Growth of Australia: The shaping of a country.*

An essay in this collection, "Trees, Trees, Trees", won the Greening Australia Journalism Award and the Landcare Media Award of 1990 after publication in the *Independent Monthly* as "The Billion Trees of Man".

Eric Rolls was made a Member of the Order of Australia for services to literature and environmental awareness on Australia Day 1992. He is a Fellow of the Australian Academy of the Humanities.

Pier Book Exchange
572B The Esplanade
Urangan QLD 4655
Ph: (07) 41289533

Other books by Eric Rolls

Sheaf Tosser
They All Ran Wild
Running Wild
The River
The Green Mosaic
Miss Strawberry Verses
A Million Wild Acres
Celebration of the Senses
Doorways: A Year of the Cumberdeen Diaries
Selected Poems
Sojourners

Eric Rolls
FROM FOREST TO SEA

Australia's Changing Environment

University of Queensland Press

First published 1993 by University of Queensland Press
Box 42, St Lucia, Queensland 4067 Australia

© Eric Rolls 1993

This book is copyright. Apart from any fair dealing
for the purposes of private study, research, criticism
or review, as permitted under the Copyright Act, no
part may be reproduced by any process without written
permission. Enquiries should be made to the publisher.

Typeset by University of Queensland Press
Printed in Australia by McPherson's Printing Group

Distributed in the USA and Canada by
International Specialized Book Services, Inc.,
5804 N.E. Hassalo Street, Portland, Oregon 97213-3640

Cataloguing in Publication Data
National Library of Australia

Rolls, Eric C. (Eric Charles), 1923– .
  From Forest to Sea

  1. Environmental protection — Australia. 2. Natural resources —
Australia. 3. Man — Influence on nature — Australia. I. Title.

363.700994

ISBN 0 7022 2576 2

For our grandchildren, aged in order from twelve years to twelve months (October 1993) — Nicola, Philippa, Emma, Hamish, Sophie, Olivia, Dominic — all of whom, if the right work is done, will bring up their own children in an even more fascinating world than we know now

Contents

Acknowledgments *viii*
Out-of-the-way Australia *1*
The new beach *10*
Country and city *18*
Spring *26*
The soil that gives me substance *31*
The song of the dingo *44*
Trees trees trees *54*
A voyage of discovery *83*
The blue-green Darling River *101*
The north coast story *125*
More a new planet than a new continent *161*

Acknowledgments

Some of these pieces have been published before: "Out-of-the-way Australia" appeared as "White Man's Dreaming" (*Independent Monthly*, July 1989); "The new beach" (*Age*, 1984); "Country and City" (*Age*, 1987); "Spring" (*The Four Seasons*, Angus & Robertson, 1989); "The soil that gives me substance" (*Homeland*, ed. George Papaellinas, Allen & Unwin, 1991); "The song of the dingo" appeared as "The Disappearing Howl (*Independent Monthly*, February 1992); "Trees trees trees" appeared as "The Billion Trees of Man" (*Independent Monthly*, June 1990); "A voyage of discovery" appeared as "Turtle Ho!" (*Independent Monthly*, September 1992); "The blue-green Darling River" appeared as "The River" (*Independent Monthly*, December 1992); and "More a new planet than a new continent" (*Essays in Environmental History*, Oxford University Press, 1993).

"The new beach" was written before the death of Eric Rolls's first wife, Joan, in 1985. In the other chapters he was accompanied by his wife, Elaine van Kempen.

Out-of-the-way Australia

Australia seldom flaunts its marvels. Unless a traveller knows what to look for, he can drive past unaware that they are there. It is the aloofness of Australia that enthrals me. It shows its wonders once in a lifetime.

In the south-west corner of Ireland, parallel ranges of hills run down to the sea. Heavy clouds come in quickly with the wind behind them and they collect on the hills for somewhere to rest. Trees and rocks grab at their undersides and they begin to roll. Soon there are lines of clouds where the hills were and green sunlight pours down between them. The wind drives the clouds up into the northern mountains where they collect so much moisture it can no longer hold them up and they begin to tumble back down against it. This time they are too heavy to ride the hills. They flow down the valleys, filling them with rain while the hills warm their rocks in bright sun. If you go to the County of Kerry will you see this marvel of weather? Yes, if you stay a day or two, or a week or two if the weather is unkind.

In the Tanami Desert in Central Australia the clouds perform over a flat plain mauve with the interaction of red soil and grey-green spinifex. They come as a huge, irregular mass, black and purple and angry grey, from anywhere out of a clear, blue sky. Lightning crazes it, then suddenly, as though the lightning has really cracked it, it divides into separate storms, half a dozen or more, that race off across the desert in different directions. One hopes that there is rain in them. If lightning starts fires, it is a dangerous place to be. Sometimes two storms pace one another 100 metres apart. They can soak two parallel strips 5 kilometres long, 200 metres wide, while the yellow sun streams between them. If you drive north-west of Alice Springs along the haphazard tracks known as the Tanami Road will you see that marvel of weather? If you go there for one month a year for twenty-five years, perhaps you will, perhaps you will not.

We came out of the Northern Territory on the Plenty Highway and headed east for Winton in the central west of Queensland. A notice on the border warned there was no food, no water, no fuel for 300 kilometres. There was no road either for long stretches, merely bands of tracks 100 metres wide where drivers had tried to skirt dust or mud or rocks. We were anxious to cross the Georgina Channels. Rain was coming down from the north, up from the south. It was reportedly two days away, but if it met early we would be cut off for weeks. At the first dry channel a dinghy was tied to a Coolibah on the top of the bank. The country had been five years in drought. How long since the boat was used? It looked as preposterous as the boat Sturt carried up Depot Creek.

Two hundred kilometres out of Winton, outliers of the MacDonnell Ranges run parallel to the road about 20 kilometres to the south. It was still hot afternoon, storm clouds had built over the range and the light was shifting weirdly. Cloud became hill, hill became cloud, both lost their form and became blank horizon. A paddock, clearly an open paddock, with sheep grazing in it shivered and grew a hill — or was it cloud? What was that long object that crossed the road in front of us? It looked too steep to drive over, and as we approached it seemed to grow more solid. Had there been a landslide? Did the road disappear under rocks? We slowed down nervously and edged into cloud so black and dense we could touch it.

On a hot day, a really hot day, drive the blacksoil plains between Baradine and Coonamble, or Narrabri and Moree, in north-west New South Wales. Eagles or hawks of a dozen species perch on the telegraph poles and supervise the plains. Nothing moves in the air or on the ground without them seeing. A Kestrel will dive for a grasshopper, a Wedge-tailed Eagle for a hare. The earth curves unbroken to the horizon. These plains are the domain of Blow-away Rolypoly, of clumps of Weeping Myall, of mirage. The distance is lost in apparent water. One can see the waves. Approaching cars look like double-decker buses, they bear their images on top. Windmills turn with their inverted forms spinning immediately above them and, since they are out of the waves of heat, the images look the more solid. Sometimes the reflections are not inverted. I have seen a whole paddock of cattle grazing normally four axe handles above the horizon.

Flying from Adelaide to Perth in the late afternoon, the plane almost keeps up with the setting sun. One interferes with time. And at 30 000 feet there is the southern coast just as I drew it on a schoolboy map. Even the sharp blue edge of the ocean fades into the depths as my crayon shaded it. Shall I write in the names of the bays and the headlands? On Yorke Peninsula I could see the jetties I fished from years ago after I had given a series of Commonwealth Literary Fund talks to school children. When the Tommy Ruff were on, those fishing in the cold late night could never answer the question, "how are they biting?" They all had their mouths full of refrigerated maggots warming for the next bait-up. A lively bait attracts more fish.

But from the plane I could not hear the racket of the stonepickers at work on ploughed barley paddocks. The din of these machines is awe-inspiring and the supply of stones seems never-ending. The first working of each year turns up more. This long strip of land is only about 40 kilometres wide. The soil is low in nitrogen; winds carry moisture across it from east and west, so the barley ripens slowly into big plump grains that are superb for malting. Only that grown under similar conditions in Ireland to make Guinness stout is equal to it. The heads of barley shatter when they are ripe, so to protect the yield the Peninsula farmers roll the crops about a week before they are fully ripe, then they use crop lifters that slide along the ground and lift the straw up into the combs of their headers. That is why the paddocks have to be free of stones.

In the days of windjammers and horse and bullock wagons, Wallaroo on Spencer Gulf was the export town for barley. Hundreds of thousands of bags were loaded there. We stayed there one night in the middle of a mouse plague, a true plague — they were in millions. The bitumen road leading to the town was furred with squashed bodies. We stayed at a motel. In the morning there was a half-halo of clipped hair on each of our pillows. Mice had nibbled us during the night.

The best things to be seen in Australia do not fit personal schedules. If the time is right the land shows them. If you are there, you may see them. You may not go in expectation at a time of your choosing. Australia's resurrection plants are one of the world's marvels. After five or six months of heavy drought I can find some 100 metres from our back gate. Our farm joins the Merriwindi section of the Pilliga forests out of Baradine in north-west New South Wales. Carry in a bucket of water — 20 litres is plenty — and find a Rock Fern. These common little dark green, dry-country ferns grow 30-40 centimetres high. After months without moisture they are unrecognisable as ferns: they are clumps of shiny, chocolate-coloured, brittle bare stems with the centre stems showing a few shrivelled light green balls attached to them. Touch them. They turn to green dust. There seems to be no life there at all. The ground about will be littered with dead twigs, a few brown eucalyptus leaves, dry needles of White Cypress Pine. Probably there will be what was a clump of moss. It is still greenish but it, too, crumbles at a touch. Even with a magnifying glass one cannot detect any sign of living plants. Pour the bucket of water over the fern. It will dampen an irregular patch about 40 centimetres wide. Walk away, and come back in twenty minutes. From a distance of 30 metres the patch will look green. Scallop-edged thalli of a moss, 2 centimetres long, one broad, will lie flat on the ground and close enough to touch one another. The clump of moss will be a healthy plushy cushion with a pile 1.5 centimetres high. Five hours later it will have set spores. The other moss will already have distributed its spores and shrivelled again. It will be difficult to find any trace of it. Another longer-lived moss with dark green thalli in the form of fans will have taken its place. By next morning the fern will have come to life. Every little ball of green dust will have unfurled into a living frond. It seems exactly as though blood has flowed into a mummy.

We take our fishing gear with us almost everywhere we go. Our long rods, live worms, mussels and yabbies came with us to Coopers Creek when we went up there through Bourke, Wanaaring and Tibooburra in the far north-west of New South Wales on a remarkable trip in 1966. I was writing *They All Ran Wild* and I wanted to see goat-catchers, brumby-shooters, doggers and rabbiters at work. Coopers Creek spends years as a series of well-spaced milky waterholes, then it comes down in a slow brown flood up to 50 kilometres wide. The soil has time to rebuild itself between each irrigation and for a few months pastures that usually carry a beast to 40 hectares will fatten four beasts to the hectare. Native Spinach springs up out of bare sand and grows 60 centimetres in a fortnight; herbs, legumes, grasses mass among it. "What this country could do if it had water!" say the lovers of irrigation who last century, and this century, even considered bringing it over the Great Dividing Range from eastern rivers. Those flood plains exhaust their wondrous fertility in that one spring, then they need years to loosen up and to manufacture chemicals that cannot be supplied from a bag. Regular water would reduce them to a compressed, sulky mass.

The roads have improved since we were out there, but the country still demands great care. A four-wheel drive in excellent condition is essential, together with spare parts and the knowledge to use them; so is a long-distance radio, a winch, extra fuel, plenty of water, a six-week supply of food since only 20 millimetres of rain can close roads for that length of time, and a compass. It is necessary to read the compass, to know what direction one has come from and in what direction one is going. The tracks might look obvious, they might look like good roads. A two-day blow can move enough sand to wipe the surface clean. There is not even a suggestion on the ground of where one is going or where one has been. Fences — there are fences as landmarks — are 150 kilometres apart. Five millimetres of rain on a claypan covers tracks with a reflection of the sky. One comes over a sandhill and the tracks off its base seem to lead across a mirror. Some claypans are 15 kilometres wide. One can either skirt the encircling sandhills to find out where the track comes out or else wait for the sun to lift the water. Very little soaks in.

The landholders do not welcome travellers; they are aware of what a nuisance the ill-prepared can be. A broken-down vehicle and a driver with no mechanical knowledge could well necessitate a drive of 200 kilometres to find out what is wrong, a return of that distance to pick up a part from the workshop, then another 400 kilometres there and back to fix it. But it is both wise and courteous to advise all homesteads that one is passing through. For their peace of mind they report your progress to neighbours by two-way radio.

That country is spectacular. Bright red sandhills run parallel to one another across white clay, sometimes a white sandhill blazes across red clay. It looks as though it must have bubbled up out of the ground. The sides to the wind slope gently, lee sides drop abruptly. Bases are solid, tops active. They used to be anchored by acacias and other shrubs but rabbits eradicated most of them. Even so, in a good spring the whole land still looks like a well laid-out garden. Purple, blue and orange bushes climb the sandhills, the flats glow yellow, yellow, yellow with daisies. Deserts favour yellow. The first litters of the first rabbits that ventured into the very dry country selected for sandy-coloured fur as soon as they began to breed.

We caught good fish in the big waterhole at Innamincka where Burke established his disastrous depot, but not with the bait we had tended so carefully. After three hours without a bite I shot one of the Corellas screaming in the Coolibahs and we baited up with bloody flesh. Fat Yellowbelly bit immediately. They are unusually silver in those waters.

John King, the survivor of the Burke and Wills expedition, left a daughter on the Cooper. For years she worked as a maid at Nappamerrie, the neighbouring station to Innamincka. I wrote a poem about her.

> Yellow Alice, King's daughter, knew the milky Cooper water,
> Swept the floors at Nappamerrie,
> Cook's offsider, made the beds,
> Asked her mother who was her father.
> Mother didn't know much either:
> "Come from some place somewhere some time,
> Useless bugger, had to feed him.
> Both his cranky mates were dead.
> Sad-faced bugger couldn't sleep,
> Sat beside the fire and shivered.
> Once or twice I warmed him up."
> Yellow Alice, lonely Alice, sometimes warmed the Chinese cook.

We took our long rods to the upper waters of the Mary River to catch Barramundi and they were as useless there as our bait on the Cooper. One turns into the river just before the bridge on the Pine Creek–Jabiru road in the Northern Territory. It is fascinating fishing there towards the end of the Dry. A fresh or two will have stirred the fish up but the creeks running into the river will be fordable so one can walk a good stretch of it. The Sorghum grass on the bank is not yet too high and dense to push through comfortably. The river itself will be a metre's width of running water that connects deep pools 3 metres long, 100 metres long. The banks are overgrown with trees and shrubs. One needs a short rod with a stiff tip to cast a heavy lure under them or through them 5–15 metres into the square metre of water not overhung with branches. A 6 kilometre walk will produce two or three fish weighing 2–3 kilograms.

A bait of worms in the bigger pools attracts Eel-Tailed Catfish and the strange Fork-Tailed Catfish with a bony extension from the head halfway down the back. Both of them, also, are superb eating. The male Fork-Tail incubates the eggs he has fertilised in his mouth, up to 120 of them 14 millimetres in diameter, an extreme mouthful. Elaine, my wife, tried a small catfish as bait to lure a Barramundi out of a big pool. A young Freshwater Crocodile chopped it in half. Sometimes young crocodiles get hooked. They come out very agile and very angry. Perhaps those fishermen who fish regularly for Hairtail with fearful teeth in fearful mouths would have no difficulty in handling them, but for most it is a disquieting business clamping those jaws shut so that the line can be cut off short. The hook will disappear in days, it will cause no lasting hurt. One watches behind for buffaloes coming down to drink. If one stands up quietly and makes oneself known, cows and calves take a hard inquisitive look, then gallop away. A lone bull can be unpredictable and dangerous. His temper is fragile. If it breaks, choose a climbable tree.

The MacDonnell Range in the Centre is not a conventional line of mountains. It is more broken, more fluted, more jagged, smoother, flatter, more equipped with outliers and outriggers than any other. And in the pink desert light, according to time and to distance, it is mauve, pink, purple, red, brown, grey, green, yellow, silver, spotted, tabby, polished amber, but never blue. East of Alice Springs it crumbles into more usual shapes, and 111 kilometres out creek beds and reefs carried gold at Arltunga. A prospector found it in 1887 but diggers rushed past it to reported rubies in the Hale River farther east. They dug out magnificent red stones, hailed in London as equal to the world's finest rubies. The first consignment brought prices to match that judgement. Then the market learnt that tons of such stones were on the way. It made a panicky reassessment and downgraded them as garnets. A difference in alumina content, not in beauty, allowed the reassessment. So disappointed miners came back to Arltunga.

It was not a big field, but 100 to 400 miners worked it for about fifty years. They built huts, temporary and semi-permanent, of hessian and corrugated iron, of local stone; they dug wells. The government put in a stamp mill, a cyanide works, an assay shed, a police station. Nothing was on a grand scale, but because of its position, because of the fact that gold being there at all made the place as outlandish as Lasseter's Reef, the relics seemed worth saving. In 1981 Peter Forrest reported on the field for the National Trust in the Northern Territory. In conjunction with the Conservation Commission it made the imaginative decision to stablilise the ruins, not to restore them. The visitor can build from what he sees. Richard Allom, a Brisbane architect with a deep knowledge of old buildings, came in as consultant. He put free-standing parasol roofs of corrugated iron over some relic walls. They would recognise what covered them and be grateful for it. When the original iron was taken for other shelters, rain began to wash out the simple mortar of local mud. Other walls in better condition could be capped with stone and silicon mortar to keep out moisture.

Two rangers now care for the area. Because of its isolation, because there is no modern imposition, it is easier to look about and people it with miners than it is at Ballarat, or Beechworth or Gulgong, where they worked in thousands.

It is a singular pleasure to drive north up Queensland's coast. Once out of the concrete abominations of the Gold Coast and the steel and glass of Noosa, no one has had enough money to exploit his bad taste until one gets to Christopher Skase's circus at Port Douglas. In between, towns have an individuality rare in Australia. Tropical shrubs and vines splash them pink and red, purple and yellow. Little electric engines fuss along streets hauling goods from factory to factory on what look like toy tracks.

Townsville is a remarkable city. The Bohle River and Ross River and several creeks split it into irregular segments, Castle Hill dominates it. A rocky sculpture rising 286 metres, it occupies a square kilometre in the centre of the city. It looks as though it was erected there to relieve the flatness.

Many of the houses are old-style Queensland, rearing off 2 metre stumps to their crowns of decorated galvanised iron ventilators. Each one different, they are a special feature of Townsville. The salt marshes on the town common that is now an environmental park attract Brolgas by the thousand to feed and breed and dance. This is now the only place where this marvellous bird displays in such numbers. As graceful as ballerinas, they perform as meticulously. A young bird who gets out of step is pecked back into place by its elders.

About 10 000 Torres Strait Islanders live in Townsville. They are such a fine-looking race they give the city an air, not of difference, but of distinction.

It is a wonder to drive up from the dry mauve light of the Centre to the wet green light of Darwin. One drives from where plants and animals are conditioned to excessive dry to where they are conditioned to excessive wet. About Pine Creek the termite mounds grow taller, the leaves of trees grow broader. The land loses its smell of eucalyptus, it develops an Asian tang. At Alice Springs eucalypts droop thickly waxed leaves edge on to the sun, at Darwin they present them outspread, soft as the leaves of waterlilies. I measured the leaves on one sapling. They were 17 centimetres broad, 27 centimetres long. Some species of eucalypt are deciduous. It was simpler to drop leaves in the Dry than to make the adjustments southern eucalypts had to.

The people of Darwin are cosmopolitan and natural. The new money is Greek, the old money Chinese. A man's ability counts for him, his skin colour is of no consequence. A white girl stands on a street corner talking fluently to an Aboriginal man in his own language. Chinese and Europeans in important positions are married to Tiwi girls from Bathurst and Melville Islands. Two mayors have been Chinese and over twenty races respected them for their personality as well as their position. Japanese, Vietnamese, Thais, Malaysians, Indonesians, Chinese grow superb fruit in their backyards and offer it for sale in weekend markets. Aboriginal artists hang in the museum alongside Fred Williams, Margaret Preston, Arthur Boyd, Olsen, Drysdale, and their brush strokes have equal authority.

Australia is at once the oldest country on earth and the newest. And it can never be assessed lightly. It need not be obvious which is which. A forest that appears primeval might be little more than a century old.

The new beach

We live our days between two beams of light. It is winter as I begin this story. At late sunrise a long ray from the north-east reflects off water on to the house. At early sunset a long ray reflected off water spears at us from the north-west. The length of the rays and the angle at which they strike us vary with the tide. And day after day westerly winds channel down Tilligery Creek, the wide salt arm of Port Stephens that the house faces. Little waves, capped with little white splashes, run through the lines of oyster racks in shallow water, slap among the mangroves, then exhaust themselves against the heavy wide cast-out conveyor belt maintaining the built-up garden.

The constant motion churns up mud from the weed beds. The water grows brown. In front of the big oyster shed built a few years go, a dug-out bay for the aluminium barge and a short triangular breakwater have altered the natural currents. The water swirls a little, turns back and pauses. It takes up mud as it swirls, deposits heavy sand as it steadies. And a new beach is forming. Already soldier crabs have moved in to prove it, and sometimes at night stingrays that have finished feeding rest on it as a clean bed. At low tide next day kite-shaped depressions in the sand mark where each has lain, eighteen one morning up to 70 centimetres across and spaced half a metre or so apart. The sand looks mysterious, bearing such evidence of a meeting of monsters. They were so close to us, a few steps from the edge of the garden.

Australia's beaches are half-moons of sand framed by granite, long arcs of sand transected by basalt, straight yellow borders to crumbling sandstone, or 100 kilometre stretches of moving white sand restrained from blowing inland by parallel dunes. Some of the sand comes from shattered coastal rocks, most of it comes down flooded rivers. Waves disperse the mud and support the sand in the shaken water. Currents distribute it. Few beaches are stable. Wind and wave chop into the sand, lift it up. The currents bear old sand along with new sand. They are fast enough to hold it in suspension until something slows the water — headland, artificial breakwater, or counter-current. Then it dumps its load.

Hundreds of tonnes of sand a day sweep through the heads of Port Stephens with the flooding tides, tour down the bay, turn and sweep out again on the ebb. Some sand adds itself to the already big sandbanks in the harbour, a little travels on up the arms. That is where the sand comes from to make the new beach.

Joan and I are living at this oyster farm north of Newcastle for some months while the owners, friends of ours, drive around Australia. Joan leaves often because pharmacists ring up for her to relieve them. I sit before a non-distracting blank wall and live what I am writing. For hours at a time I move between nineteenth-century China and Australia. The room is full of musical Cantonese dialects. A gambler on a second floor lowers money in a bowl on a string and chants his bets. He holds on to the end tone, it rings round the room. I've done enough for a while. I open the door and go out. And 400 metres across the water, from a mangrove on an island, a White-breasted Sea Eagle picks up the note I was hearing in imagination and swells it to an immense real cry. Again and again he challenges kilometres of water.

Groups of Eastern Curlews meet on mud spits and whistle plaintively through their long curved bills, or chatter harshly. Sometimes they cry onomatopaeically "Kerloo! Kerloo! Kerloo!", telling us what they are. It is a ghostly cry at night. And there are ghosts here. The house is built on an Aboriginal midden to lift it above the sour salt marshes. Tribe after tribe came from inland for yearly feasts on Sydney Rock Oysters. Listen! You can hear 50 thousand years of people laughing and exclaiming over superb food, then the rattle of empty shells thrown on the pile.

Joan and I eat oysters too, a dozen each for lunch most days, plump, delicious, full of creamy spawn. A couple of times we opened fourteen dozen small ones, a jugful of oysters, poached them in their own juice, chopped them, added milk, cream, grated onion and egg yolk in a double saucepan and made a bisque, the richest dish in all the thousands of years of feasting.

It is a vital seam, the joining of land and sea. A sharp-winged Common Tern, head down, patrols water. Two metres from it, a sharp-winged Brown Falcon paces the tern, wing beat for wing beat, and patrols land. But not all creatures keep themselves apart. The tidal zone is a banquet spread for land and sea.

As the tide comes in, repulsive, poisonous toadfish suck up what the first water floats to the surface. They slurp, make rings with mouth and tail. Schools of poddy mullet follow, thin brown shapes glinting silver when they turn. They glide over mud and sand, feeding quickly, changing place back to front like galahs. Sometimes they mill over something, then settle down to feed in the shape of a many-spoked wheel. Yellow-finned bream and little whiting and a few other species follow the mullet. When there is 60 centimetres of water over the weed bed, the big fish feed. Whiting patrol in fast schools. Flathead settle down individually on mud or sand and wait for food to drift over them. Bream examine the oyster racks, and mullet begin to jump, skittering over the surface when something chases them.

As the tide goes out, crabs come up to see what it has left, so many I can hear them: a rustling, a scraping of mud. They are nervous. They do not feed in wind or rain. A blown leaf freezes them, a swallow diving in play sends them scuttling sideways to their holes. My shadow puts them right down their holes for half an hour or so. They can see me from 30 metres. I crawl to the garden's edge, lie down and watch. Crabs of all sizes come up, according to age, out of pinholes, out of holes 5 millimetres in diameter, 10, 15, 20 millimetres, spaced a few centimetres apart. All begin feeding at once. There seems no end to what the tide leaves. As one claw rises to a mouth the other picks something up, over and over without hesitation. They move a few centimetres, begin again.

I never saw the soldier crabs on the new beach, only the bubbles of sand where they had fed. It was a colony of young ones that feed just below the surface, rolling up sand as they sift it for food and pushing it out of the way.

The crabs that live among the mangroves are obvious. The mud flowers with their short, bright, pink claws. Sometimes one stops feeding and has a brief altercation with a fast-moving, long-clawed crab that runs about among them. Away from the mangroves drab brown crabs harvest the weed beds. I dig them up when I turn over the soft mud collecting worms for bait. The worms are so interesting I sometimes feel dubious about taking them, and I remind myself that my pronged hoe is no better a gatherer than the beak of a stingray or a heron. One bloodworm is a sequence of thick, soft segments, so fragile it builds a tube to live in. The tube is astonishingly sturdy and glossy smooth on the inside. The worm is so delicate it never leaves the tube. To feed it brings its head to the entrance and drifts a ring of long, hair-like, fleshy collectors into the water.

Birds come to harvest the seething life: curlews, sandpipers, herons, ducks, egrets, ibis. Short straight beaks, long curved beaks, thick bills probe, lift, swallow. Hundreds of thousands of living things disappear in a haphazard execution. Nobody's God supervises each mouthful.

And oyster farmers move to work their exposed racks. We hear them coming back, a knock knock above the many engine notes as they cull the oysters in the black line of towed barges. With a glove on the left hand, an old file in the right, men tear off sticks fruited with oysters and knock them off into bins.

Some nights we hear the surf from the oyster farm. Five kilometres of trees and high sand dunes smooth out the surges so that the noise comes to us as a continuous, muted drone. I rode over those sand dunes during the early days of World War II with the 24th Light Horse. I was 17. We had been in a mock battle with an infantry battalion in the then extensive scrub about Port Stephens. We never engaged the enemy. Perhaps they had been shipped overseas before we arrived, though one day we got a fright. The guard might have heard a mob of wallabies in the bush. Anyway, he called an alarm. We were sitting down eating cut lunches. Our horses were tied to trees with neck ropes, nosebags on, girths loosened. We abandoned our sandwiches, raced to our horses, untied them, slipped nosebags down their necks (a vital move, no matter how close the enemy, else the excited horses might breathe in chaff and choke), leapt on and galloped away, girths still loose, bits dangling and double reins and neck ropes caught up anyhow in the left hand. At least we had our empty rifles at the ready in the right. The horses took thorough offence about it, and we had to talk to them calmingly every time we approached them for days after.

At the end of the stunt we rode out of the bush and over those sandhills towards the sea in Newcastle Bight. The series of high, loose, parallel dunes is 3 kilometres wide in places. The horses floundered up and down them, bogging to their knees. The smooth sand churned into humps and hollows, but immediately the wind levelled it. Three metres behind each horse there was no record of its passage.

On the beach we unsaddled, undressed, then a regiment of naked men rode laughing into the surf. Our horses were lively and sturdy, the waler type. None of them had ever seen the sea. They snorted, pricked their ears, humped their backs when water lapped their bellies. Once we had coaxed them into the surf they loved it, meeting the waves breast on, swimming out to the next. When my mount had had enough I allowed him to turn back towards the beach. He began to whinny with excitement. We came out on to hard sand. The excitement had been too much. He lost his good manners and bucked. I was wet and slippery, the horse was wet and slipperier, a greasy pole. I came down hard. Then other horses began to buck. Men thudded off all along the beach and fifty horses raced for the sandhills. It took hours for the half-dozen men who had stayed mounted in their saddles for just such a contingency to round them all up.

Now, off and on, I fish that beach, walking along it in the late afternoon looking for channels alongside sandbeds — they are always shifting — and marking each good spot with a square of newspaper high on the beach so I can find it at night. Some kilometres of the north end of the beach were closed for several years. The RAAF used beach and dunes for bombing practice. I was at Nelson Bay on Port Stephens when the unexploded bombs were cleared and the barbed wire barricades taken down, and was one of the first to walk the beach again. Pipis had come back — over-harvesting had thinned them on most beaches. Not only were they on display as paired holes in the sand or ripples in the water, colonies of them were coming in on the waves, all as big as those I had read of that were harvested for canning by steam dredges in the 1920s. Some waves looked solid with them, 30 metres of pipis a metre high. They must have means of propulsion in the water. Perhaps they flap their shells like scallops. They stayed with a wave as it broke, then allowed it to spread them on the beach. Immediately each shell opened a little, the tough foot protruded, the shell cocked on end and sank into the sand. Within seconds there were none to be seen.

I use them for bait, along with worms. It is entralling, alone on a beach at night, connected to the sea by a line. The sea talks up it: which way the currents are moving, what weed is floating, what fish are feeding. One night on a little beach known as Box Beach I caught a stingray. I suspected what it was before I saw it because the heavy weight on the line pulsed, and every now and then it sank down and needed a lot of careful lifting from different angles to get it moving again. With the help of a bigger wave I finally floated it up on to the sand. The wings are good eating but I did not know how to handle the creature. It was 60 centimetres across. How far could it twist the mouth that could certainly bite off a finger? And how did it strike with the poisonous spine on its tail?

I decided to put it back in the water and bent to cut the line. The stingray heaved. I jumped back. And there were two stingrays on the beach. It had given birth to a baby that was astonishingly big, at least 20 centimetres across. Its wings undulated as though it was trying to swim. The mother heaved again and another young one slid against the other. I pushed all three back into the surf with the handle of my gaff, keeping them together. I did not know if the mother cared for the young or not. There was enough moon to watch them swim away, black bats of the sea.

At Seal Rocks to the north of Port Stephens, a little beach used by professional fishermen faces open water, so they built their own boats on the beach to suit the conditions. Ron Legge and his father before him went out with axe and adze and cut knees for the boats out of tall, close-grained paperbarks. Keel and stem had to be in one piece for strength, so they chose trees of perfect grain with a big root coming out at the correct angle to form the bow. The boatbuilder shaped the ribs out of tough, springy Spotted Gum.

The last boat was built only three years ago, of plywood and fibreglass, but to the old pattern, 8 metres long, 3 metres wide, with a shallow 60 centimetre draught. The fishermen launch them either by pushing them on rollers into the surf, or pulling them in with an engine-driven winch and a rope running round a big cement block out in the water. The winch is on the opposite side of a gravel road to the beach so a notice warns "CAUTION MOVING WIRE ROPE ACROSS THE ROAD."

Waves break over the boats during the launching, so storage wells, the engine well, the well for the winch (to pull up the traps), the well to work the tiller are covered tightly to shed water. The boats are as flat on top as a surfboard. When it begins to float, the fisherman leaps on, lifts a cover, starts the engine, lifts another cover so he can work the tiller and heads for his craypots and fish traps up to 30 kilometres offshore. If a sea comes up while he is out — there is nowhere to put a radio to give him warning — he ties the tiller, puts all the covers back on, and rides home balancing like a surfie.

Fast aluminium boats are taking over with twin outboards for safety. They rest high on the beach and an amazing collection of old tractors pull them in and out of the water. A couple of tractors sit immobile on the sand as rusted lumps. Others, rusty enough, still work; some are brightly painted. A diesel Toyota Landcruiser pulls boats so far into the water that when it turns waves slap against its sides. The rusty body must soon crumble into red sand.

When the boats come in laden with fish and crayfish, the fishermen wait in deep water while they clean the last of their catch. Gulls scream and snatch the gut as it leaves their hands. Then, on one of the regular higher waves, the boats charge ashore.

Until the mid-1930s the Seal Rocks fishermen took their catches out by horse and cart. They smoked much of the fish over slow-burning banksia that gave it a delicious flavour, they kept crayfish alive in big netted tripods sunk into the water. The huge Australian Fur Seals that give Seal Rocks its name eat crayfish, but no one begrudges what they take. They seldom show themselves to shore watchers. From a boat they can be seen sprawled on the open side of rocky islands, great brown mounds of blubber.

Dolphins come in to play in the waves off one little beach that shelves down so quickly a wave can be 3 metres high 3 metres from dry sand. The game they play is to swim parallel with a wave as it comes in. The water is often so clear one can see them plainly. They hold themselves midway up a wave and move along it so that there is no more than a film of water between skin and air. It is an opposite game to the one surfboard riders play, charging down tubes without breaking water with head or hands.

Our beaches were vibrant with Aborigines for thousands of years, then they were quiet. They quietened suddenly about Sydney as early as 1789. Hundreds of Aborigines burning with smallpox came down to cool themselves and died on the wet sand. For more than a hundred years after most beaches were merely for white men to fish. Women seldom saw them.

I envy the catches of last century's fishermen. But I do not envy the coarse gear they fished with. It is more exciting now fishing with fine lines and reels that give such exquisite touch, even though catches are less.

And I do not envy the crude morals of last century: women giggling in bathing boxes, men bathing by night because it was obscene to bathe by day. Even in the early years of this century men and women wore more to a swimming party than they now wear to town. I love the garish beaches of present day: flowering with umbrellas, carpeted with towels, glinting with cans, ballooning with bellies, shining with smooth, oiled backs, happy with children, humped with buttocks, quivering with breasts, even writhing with penises on some beaches.

Australian sunshine has lightened lives. Today is a good time to be alive.

Country and city

Yesterday Elaine and I drove back from Sydney to Baradine. Where did we cross the divide between city and country: Pennant Hills, Kurrajong, Lithgow? The federal government had no doubt after an outbreak of smallpox in 1913: it was the circumference of a circle of radius 25 kilometres from the GPO. To move over it without a medical certificate declaring one free of smallpox meant gaol.

What did we leave behind and what did we drive into?

This morning our late spring birds woke us gently. Half an hour before any other bird called, while it was still so grey he must have read the time from the stars, a Rufous Songlark throbbed a quick succession of notes, then chirred and whistled and sang so diversely he might have been several birds. He sits on the bare electric wire for that regular broadcast. Five thousand volts run between his feet.

I opened my eyes again as the light was turning golden and a Brown Songlark was putting on his act for his five anonymously-brown females. He is one of the few birds who keeps a harem. As they search the ground for insects, he flutters up, flutters down, hovers and sings. All his movements are so jerky he seems powered by faulty clockwork. I did not see him, I heard him. His powerful voice sounds as though it might once have been tuneful. Now it comes out of a scratchy music box in a tin bird.

A Grey Shrike-thrush sang three of his airs. He does not sing notes, he sings harmonic chords, and makes a Nightingale sound insignificant. I got up when the pair of Willy Wagtails guarding the three nestlings that took to the air while we were away began to scold the first Currawong to fly in from the forest to feed in the garden.

The bedroom door was wide open, the blind up on the big window.

I walked naked into the kitchen to set the fire. Every window in the house was open, every door. There are eight doors opening to the outside and not a key to fit any of them. We have been here twenty years; the keys might have been lost or thrown away fifty years ago.

I light the slow combustion stove with pine chips split from a dead log on our farm. They catch with a quick lift of flame and aromatic smells. They cannot be used to start open fires, they spit hot coals 2 or 3 metres, but the firebox contains them. The wood that sustains the fire all day comes from 300-year-old ironbarks that died in the 1902 drought. There is no sap left in it to damp it down sulkily. It glows. Through the kitchen window drifts the heavy scent of big White Cedars mauve with blossom. A Silky Oak is an orange cone of flowers 15 metres high, and raucous honeyeaters have begun to squabble over the nectar that formed during the night. They have fleshy bristles on the ends of their tongues to brush it out. I can hear no vehicles, no strange voices. The corrugated iron roof creaks as the sun warms it. If a cloud covers the sun, it creaks a different note as it shrinks a little. It responds to rain by drumming, gurgling, rippling, ringing. It is more than a cover, it is a sensitive skin.

No one expects us, no timetable orders us, we need not hurry breakfast. There is the yoghurt we make, the muesli, the wholegrain bread Elaine bakes, eggs from hens that run free several hours a day, the liver of the last lamb I killed or the last chicken, grapefruit and oranges from our own trees. Tomatoes finish their ripening in a line along the window sill. Until two or three years ago I always milked a cow or two and I think now of beginning again. The clotted cream skimmed from bowls of milk is the finest product of the soil. From separated cream, that extracted from milk whirled through the fins inside a spinning, hand-turned cylinder, I made butter. Stir cream steadily in a bowl with a wooden paddle. It thickens, breaks, and a liquid slops about a honey-coloured solid that tastes richer than factory butter ever tastes. I always felt I stirred sunlight through it.

It is the dignity of life in the country that now most distinguishes city and country living. With no immediate neighbours, no one else's buildings to throw shadows or blot out a view, little mechanical noise, no imported sparrows quarrelling in gutters or Indian Mynas squawking in someone's backyard tree, no traffic lights between home and town — probably not even a car — no dog turds on footpaths, no street lights to be shaded out or thieves to be locked out or strangers to be accommodated by turning the radio down late at night or regulating the tap so that the hose does not spray the footpath: by escaping the obligations and obtrusions of city living, we can be more responsible for our own condition.

And we no longer have to make sacrifices to lead such a remarkable life. Twenty years ago one had to make a difficult balance of independence and intellect, experience and excitement. We did our best but too often we missed films or plays that generated new thought, we missed recitals, we missed the conversation of people expert in other ways. There is no doubt that the cities produce greater intelligences. My parents realised that fifty years ago when they sent me to compete with the best at Fort Street Boys' High School.

Now both economic conditions and new inventions have linked country and city. Businesspeople have moved on to the land, either permanently to manage farms to make money or to experiment with country life in weekend hobby farms. Farmers, some of them incompetent, some of them the victims of extravagant and inexpert advice, have moved to cities to find work. Two cultures are merging. My children were born on the land, they absorbed its experiences deeply, they went to school in Sydney, to university. They worked in cities and overseas as physiotherapist, veterinary scientist, air traffic controller. Now two of them are back on the land.

My daughter, Kerry Jane, can drive 5 kilometres to put her daughter, Nicola, on the school bus in the morning. She can strap Philippa in their aeroplane at the back gate, fly to Mascot, land on the tail of one jumbo jet, swing out of the way of the next one coming in, park her plane, pick up urgent machinery parts temporarily out of stock in the country, and fly home again in plenty of time to drive out to pick up Nicola from the school bus. Their farm is connected to the world by telephone, by telex, by fax.

It is over 500 kilometres from Baradine to Sydney. Most local farmers go down so often they can say within fifteen or twenty minutes how long it takes to drive down. And they are as expert in Sydney traffic as on a slippery blacksoil track.

There are still misconceptions and disconcerting inaptitudes between city and country, country and city. Joh's [Bjelke-Petersen's] rural maxims sound years out of date: "I wouldn't throw him a barbed wire to pull him out of a bog." A recent issue of the *Land* newspaper carried a recipe for Oysters Kilpatrick, a savage dish that an expensive city restaurant might well have served in the 1950s. The CWA has not adjusted its prurient moral attitude since it was formed. A discussion about Aborigines with Western Australian landholders or the rednecks of north Queensland sets one back fifty years. Our local politicians can sometimes be worse than embarrassing. "I've swum in dieldrin," said Wal Murray at the height of the recent consternation over contaminated meat. "It never did me any harm." Our meat exports are so important that what was at stake was Australia's future.

But the city supporters of the Reverend Fred Nile are more curiously obsessed with sex than any member of the CWA. To walk into a dockside pub at 6.00 a.m. when the night shift of wharfies is coming off is to walk into last century. The bosses are presented as the drivers, the workers as the driven. The talk negates the history of trade unions. "What do you think of the Mudginberri dispute?" asked a beef skinner at a country abattoir. He held his knife next to the hock of a steer especially fattened for the Japanese market. It was worth over a thousand dollars. The owner was standing nearby. "Buggered if I know," he answered. "Don't know anything about it. The men might be right."

"Good thing you answered that way," said the abattoir worker, "or I'd have ripped the back fat right out of him." There is little understanding of one man putting up his money and management, the other his manual skills, for the same purpose. Some of our left-wing city Labor politicians are as out of touch with modern times as the crudest rural Queensland National Party representatives.

Arthur Hoey Davis — Steele Rudd — who published his *Selection* stories early this century, erected a comic barrier, a dog-leg fence between city and country. He did not intend to. What he first wrote was serious. He told of men and women with too little money, too little land and too little knowledge trying to cope. He had had bitter experience on his father's farm.

Sandy's selection was on Sleepy Creek, a hundred-and-sixty-acre one. All beautiful land, too. Three acres of it were cleared. Sandy was struggling hard to clear one more. The remaining hundred and fifty-six were under scrub, prickly-pear, wallaby-bush and Bathurst-burr. He was ploughing and Jimmy tried to help." Sandy continued to yell at the horses ... "To y'! To y'!" Sandy called when the black mare left the furrow. Jimmy pulled the wrong rein. "Damn it!" Sandy shouted, losing his temper. "To y'!" Jimmy pulled the wrong rein some more; then Sandy threw the handles

down savagely, and cried, "Blarst it, what's the good o' that?" and
snatched the reins from Jimmy and nearly dragged the heads off the horses.

When city dwellers, misunderstanding the enormity of trying to
handle Australian soil with no proper equipment, found such scenes
funny, Steele Rudd set out to feed them. Instead of the relentless truth,
he presented best-selling farce that became the *Dad and Dave* of radio
and ran for years. So he changed the attitude of city to country from
mild envy to mild amusement. Australian soil can still make the best
of farmers look incompetent. It never allowed some of the settlers of
the 1860s and 1870s to make more than a beginning.

In Australia's white beginning, anybody who could grow food or
husband animals was of unusual importance. The colony was hungry
and nothing had been planned. There were few gardeners, fewer
farmers, few tools to work with. Most of the newcomers were products
of the slums who had never been in contact with soil. They were
artificial people grown on bricks and flagstones. They knew as little
about how food is produced as a modern city dweller buying packages
in a supermarket.

By the 1830s, 1840s and 1850s antagonism had grown up between
the many new farmers and graziers who had either learnt their jobs in
Australia or come with centuries of experience behind them, and the
carpenters, wheelwrights, painters, labourers, bricklayers, fell-
mongers, clerks, stonemasons and shop assistants of the cities. The
countrymen needed cheap labour to build farms out of thousands of
years of soil that had known no cultivation. When they found they
could not pay the extreme rates of the expanding 1830s, they imported
Indian coolies at one-quarter rates and spoke of bringing in Chinese.
The city workers feared for their rates of pay. By the slump of the 1840s
they feared for their very jobs. Many were out of work. Landholders
trying to rebuild their fortunes brought in Chinese labourers from
Fujian Province. There was an outcry from city workers who also
cursed frequent boatloads of poor Irish and Scottish immigrants. City
and country had opposite needs.

In the 1850s the landholders on the New South Wales Legislative
Council brought down a Gold Fields Management Bill designed, by
extreme charges, to force excited miners looking for independence
back to sober work. Many, instead of responding as was intended,
walked through to the gold and the milder laws of Victoria. Those left

on the Turon diggings in New South Wales erected a stuffed effigy of that self-seeking squatter, William Wentworth, hanged it, shot it, then blew it apart with gunpowder.

It was Sydney shopkeepers who blew the bill apart. They lost so much business with the miners who left that they drew up a petition to the Governor and collected the signatures of 6000 city residents. After a sufficient interval to show it was not coerced, the government moderated the laws.

Differences now are slighter, though the intelligent country dweller has the advantage of the city dweller in experience and understanding. Les Murray's poetic magnificence would have been stunted without his contact with the soil. I could not have grown any of my books in bitumen. I write to explain Australia so I have to see it uncovered.

The conservation movement and counter-movement present the greatest differences in outlook. Both sides are usually wrong because they know nothing of what has happened to Australian soil. Farmers want to clear more land. Two centuries of bad laws and pragmatic government have jeopardised farmers' survival. Conservationists see a tree as a sacred edifice, farmers see it as an obstruction. Wide machinery, that is to say economical machinery — demands open paddocks. If a farmer is pulling a scarifier 20 metres wide and there are two trees 19 metres apart, it is certainly sensible to take out one of them or else an unprofitable area of unworked ground has to be left around them. But if there are a few hectares of trees in a corner of a paddock, the farmer ought to fence them off and plant shrubs among them, not push them down. They could yield insects and birds that will be valuable. And the farmer's neighbours, by judicious planting or fencing, ought to make an alleyway to their own saved trees.

There are now many more trees in southern Australia than there were 200 years ago. It happened by a quirk of conditions that there is no room to explain here. But that does not make trees less important. They are inordinately important now because they have to do the job of maintaining the land that deep-rooted grasses and low-growing shrubs once held together. It annoys a farmer to be told by city conservationists that he has cut down too many trees when he knows that he has more trees than the country supported when his land was taken up, and the conservationist is often too ignorant and stupid to be able to explain to the farmer why he now has to grow more trees. They build antagonism, not good soil.

Dry-country graziers and animal welfare groups spar like pairs of old bucks over what is to be done with kangaroos. Save the Kangaroo committees are fervid, vocal and dangerous. They do not know that there are now more Red and Grey Kangaroos in Australia than there were 200 years ago; they ignore animals like the Bilby or the Mala or the Golden-shouldered Parrot that need attention; they ignore the fat-free flesh of the kangaroo and its skin which makes the finest shoe leather. The landholders laugh at the suggestion that they should learn how to farm kangarooos; they ignore the fact that their land is giving notice that it cannot tolerate sheep and cattle. Both sides are willing to forfeit the soil for different reasons.

Country children learn of sex, birth and death easily. They are constant occurrences. It gives them a great advantage since they understand deeply what can be frightening to city people with no natural experiences. From the time that they were very young, our children rode around the lambing ewes. They could recognise a ewe that was having trouble. Often they helped her. Their small hands fitted in to straighten a bent foreleg much easier than mine. Few country women have any fear about giving birth. Most of them have attended many more births than the doctor attending them. And provided they can persuade the nurses not to fill their newborn babies up with glucose, they have no worries about having no immediate milk. From their experience with animals they know it takes hungry sucking to get it flowing. Yet I have seen a city woman offended when compared with an animal. "I'm not an animal," she said angrily. What else is she?

When they were 9 and 7 years old, our children had named the Dorset Horn rams after their sexual abilities. There were Jealousy, Greedy, Clumsy, Slowcoach, Quick and many others. Children who grow up in cities seldom reach that understanding of sex as comedy, not mystery.

They learn to treat all animals with respect, with authority and with kindliness, even though many of those they handle are bred for meat. They never tease animals because they know that an insulted animal can turn vicious. A big moth or a grasshopper bashing itself against a light does not mean a reach for the spray can, it means a reach for something to catch it in, so that it can be let go outside. Wolf spiders on the floor are welcome, as are Huntsmen on the walls, but 4-year-olds search their sandpit toys for Redbacks before they handle them. A snake in the house is not a terror, though it does mean caution. The beautiful Yellow-faced Whip Snakes are no problem. Although they are poisonous they seldom strike. We roll an *Age* or *Sydney Morning Herald* from corner to corner and place it in its path and the snake seeks sanctuary from the bewildering surface it finds itself on. Then we fold over the ends of the paper and carry the snake outside. A Brown Snake is a nasty visitor. They get very angry and very dangerous, and if they cannot be persuaded to head for an open door, we have to get the little 410-gauge shotgun and blow a hole in snake and floor.

Familiarity with guns and rifles is another difference between country and city. Every household owns several, every country child learns to shoot. Forty years ago, it was custom to give a boy a pea-rifle (.22 calibre) on his twelfth birthday. There were more rabbits and less people in those days. One can no longer shoot so freely, but all still become expert shots. They know exactly what they are shooting at, why they are shooting it, and where the bullet will end up before they pull the trigger, even at night when all they can see is a pair of eyes in a spotlight. The eyes of each animal reflect light differently. Some city shooters drive hard all Friday night until they reach the western districts of New South Wales or Victoria. In two days of senseless destruction they shoot signposts, property names, mailboxes, water tanks, windmills, pigs, goats, sheep, cattle, horses and an occasional human.

Harvest is now coming up. Some farmers are already stripping oats. We work by seasons, by the weather, not by the clock. We work according to what there is to do, not by hours a day. If we have high expectations of ourselves, nobody stands in the way of our attaining them.

Whatever misunderstandings, whatever misconceptions still exist between country and city, the one vital fact is that they are now interdependent. One cannot function without the other.

Spring

Australia moves by fits and starts. Our seasons do not acknowledge 1 September. Even changing hours of daylight influence exotic plants the most. Sow tomatoes for best yield when hours are increasing, carrots when they are decreasing. Over much of Australia native plants must grow when it rains, whatever the season. They might not get another opportunity for months.

When we lived at Boggabri we planted a weeping willow at the end of the kitchen drain. It grew quickly and we decided that each year its first leaf would mark the beginning of spring. All the family watched for it. One whole leaf had to present itself to earn the glad cry "Spring is here!" because some years a bud would split and the tip of a leaf show, then the temperature would change and it dared not risk uncovering itself any more. The earliest it appeared was mid-July, the latest mid-October. At the same time — it was always after good rain — all plants sprang into sudden growth. In one week crops and pastures made a month's growth. It happens everywhere every year, some time between the cold weather and the hot. That is the spring. The word for the season has the same Germanic origin as "spring", meaning "to jump".

The lovely Chinese character for spring, shows a plant coming to life after the thaw. A seedling warmed by the sun, 日 , twists its stem, ㇄ , as it tries to push through the hard ground, ▬ . Once it breaks the surface, the cotyledons open, ᴗᴗ . These symbols all combine in the one old pictogram for spring, 萅 . Over the centuries scholars and artists refined it into this character to exhibit their brush strokes and their sense of balance: 春 .

After the regular visitation of winter snow over so much of China, spring is such a definite season it divides into six joints and breaths of a fortnight each, the Beginning of Spring, Spring Rains, the Feast of Excited Insects, the Vernal Equinox, Clear and Bright, Grain Rain.

The grain rain must come much earlier in Australia. Except in the southern districts it is usually not summer's job to ripen crops sown at the end of autumn or the beginning of winter. The principal rain must follow soon after the spring bursts, since there are so many more leaves to feed. Once it comes, flagged shoots lift up out of the wheat plants. One can draw them out easily and slide finger and thumb down their sappy lengths to feel the little bulges of the heads. They might be no more than 15 millimetres long, but one can see where the grains will form. The noded grain stem pushes the growing head through a cylinder. In our district it lifts clear of the flag in mid-spring, it flowers, it fills. The farmer's year's work is in flower, his lifetime of knowledge modified kindly or savagely by the weather. A freak late frost can destroy a crop, so can a week of blazing heat when it is in flower, but good rains until the grain is doughy, a drying-off to harden it, then hot weather for harvest can build it into magnificence: work, knowledge and plant yielding their utmost. The testing of the first grain is a farmer's spring joy. Snap off a golden head — the stem is now brittle. Grind it in the cupped left hand with the heel of the right hand. Blow away awn and husk, grind a little more, blow again. There might be sixty grains, flint hard and almost translucent. Chew them. In a minute or so a ball of gluten forms in the mouth. It is a wonder to put the header into such a crop. It brings more than profit: the soil has justified one.

In the Top End, that wondrous part of Australia so many Australians never see, there are no conventional seasons, there is a Wet and a Dry. The changeover begins some time after mid-October with a lift in humidity. The Baobab trees respond. One week their huge bulbous trunks support apparently dead branches, the next week masses of big white flowers. No leaves interfere with the display. It is a passion of flowers only. At the same time the termite mounds, 1–3 metres high, seem to break into flower. New red bulges protrude out of weathered grey pillars. Thousands of workers are making room for those that will hatch to feed on new grass when the rains come.

Lake Eyre has a spring once or twice a century when rivers and creeks of dry sand flow water for a change and pour it metres deep into its huge salt pans. Swans, pelicans, herons, gulls, fish, frogs breed there in hundreds of thousands. For a couple of years the lake is livelier than it was when it was an ancient sea. Then it dries up. Fish and frogs die in stinking masses, the fittest of the birds fly out to establish themselves somewhere else as superior representatives of their species. They have to survive a dry flight of hundreds of kilometres. I have seen three young Lake Eyre pelicans resting on one of our smaller farm ground tanks before carrying on to the coast. They looked like yachts in a backyard swimming pool. It was probably their first stop in 1200 kilometres.

The rains that spring the grasses in the Tanami Desert come in sudden storms. Black clouds build above the reds and blues and mauves of the flats and ridges, then they break up into individual storms, perhaps a dozen, that scatter across the desert in narrow paths. Creeks run for an hour or so and empty nowhere, but all along them insects hatch, or break out of chrysalises, especially moths in thousands. They have no time to feed. They mate, lay eggs and die. As adults their spring is the whole of life and it lasts a few hours of one night. Most of them have not even been given a name. They have no place in the population of the world.

As I write this, a silt trap on our biggest farm dam is busy with animals. A White-faced Heron patrols its edges in turn with a White-necked Heron. Yabbies shoot backwards from the shallows. Tadpoles and water beetles break the surface. Dragonflies and their associates whirr overhead in mating loops, then the females pull their tails free and the males circle low over the water still grasping the females behind the head with their anal claspers and dip down every now and then so that the females can trail their ovipositors in the water and lay their eggs. Only three days ago this tank had been dry for months. We had a good storm late in the evening, it filled, and by early morning yabbies had come up from watery pockets 2 or 3 metres underground; so had frogs. They had mated already and the females had laid their eggs. The black-specked froth floated in clumps. The urgency of life in Australia constantly amazes me. Apparently sterile dust becomes a busy pool of water in a few hours.

The great Murray River once knew spring excitement every year. Before the dams on its upper waters were built, snow thawing on the ranges in New South Wales and Victoria began to lift its water level about September. A few hot days, a good rain sent floodwaters down, and a succession of ducks responded. Grey Teal acted first. As soon as the banks overflowed into lagoons, flocks of males and females began patrolling them. They flew high from one lagoon to the next, they wheeled, skimmed the water, turned again. If the water kept on rising, the males formed separate flocks and began courtship displays. They spashed down into the water, trod it and thrashed their wings, bounced across it, swam in circles, lifted into the air and splashed down again. For several days the females took little notice of them. They wisely watched the water level. If it rose high enough, they selected one of the posturing males, mated and looked for a nesting site. Their young feed on plants as well as water creatures. Black Ducks waited a week of so longer for animal life to build up in the water. Hardheads nested when the waters reached out to the lignum and cumbungi swamps, then they had the deep water they prefer to feed in to fall back on. The beautiful Pink-eared Ducks with spectacular zebra stripes on neck and flanks bred after the water began to drop in lagoons. Then the water was warm enough to produce blooms of plankton for their ducklings.

Yellowbelly and Silver Perch, washed out into the lagoons, also spawned as the water temperature rose. In the next flood, adults and fingerlings washed back into the river. Now the Murray is little more than an irrigation channel. It has lost all the seasons.

Of native plants, acacias most appreciate spring. They celebrate it more or less according to the weather. I wrote a short poem about the wondrous spring of 1974 in the Pilliga Forest that forms our northern boundary.

> September comes in always yellow and raw umber.
> In the scrub acacias are in bloom in hundred acre patches.
> It is trite to write of spring and wattle blossom
> So just let me record this:
> I live where miracles are commonplace.

That spring some species of acacia bloomed in 400 hectare masses. Branches trailed on the ground with the weight of the blossom, they met overhead across the tracks and dropped flowers 3 centimetres deep. One drove through yellow tunnels. There were white wattles, cream, all the ranges of yellow in spikes and balls. The buttery yellow of cassia overlay it in patches. Purple False Sarsaparilla climbed stumps and shrubs. Angophora and bloodwood stood white with flowers. White Cypress Pine spurted pollen in heavy brown clouds. Red, blue, mauve, white shrubs of fifty species displayed in masses. Rosy pink Drosera carpeted the ground on the Broom plains. Hundreds of thousands of hectares of marvellous garden began at our back gate.

And if we had been away on research, even if I had been writing a different book then, I might have missed it. It lasted a fortnight at its best. Now conditions have changed. Trees have grown and occluded the understorey. There might never be another spring like it. In Australia one must keep constantly aware. So much that is marvellous happens once in a lifetime.

The soil that gives me substance

I interpret homeland as the soil that gives me substance and I am about to leave it for the awful authority who commands "Write this!". The soil has supported all my books; it has encouraged me with wonders. I have to leave it, I suppose, because in the end I failed it: I did not outgrow costs. The compulsion of research and writing did not allow the time to buy more land, and an area which had richly nurtured a family and a good farmhand for twenty years shrank to incompetent dimensions. So I abandon a lifetime's experience of the soil. I cannot write "knowledge" instead of "experience": one cannot know the soil. It is not an inert pudding fattening plants, it is a complex of wondrously complicated chemicals, of fungi, of visible animals, of invisible bacteria and protozoa, of minerals, elements, decaying organisms, of positive and negative ions continually adjusting one another.

Take up a small handful of soil, pick out the beetles, grubs and little spiders and the nebulous things which move, until the soil lies apparently still in the hand, then magnify it 10 thousand times. It becomes a seething mass of living creatures, of amoebae so constantly changing shape that they seem to derange their surroundings, of 150 thousand million bacteria shaped like rods, or hooks, like blobs, spots, lines, rectangles dividing, dividing, dividing and growing again so that it seems that they must fill the hand, the room, the world, until one notices the predatory protozoa dividing to keep pace with them and eating them in millions. One does not walk on solid ground, one walks on living microscopic beings which are in association and in conflict and are busier by far than anything in the world we can see.

The soil looked solid enough to me at the age of 5 when I began to take an interest in the subsoil disclosed as my father trenched the deep leads of the rabbit burrows he was digging out on a farm he had bought at Narrabri. He had netted the boundary, he had bought six crossbred dogs to hunt through the paddocks and bolt the rabbits underground where he could laboriously reach them with mattock and shovel, pull them out by the hind legs and stretch their necks. He obstinately insisted on running the deep leads to the end instead of filling them in. He liked to see his prey dead, to count the bodies, instead of hoping they were buried somewhere underground. Some of the trenches he dug were 3 metres long, 60 centimetres wide and up to 2 metres deep.

The creamy brown topsoil 30 centimetres deep was laced with the roots of wiregrasses, the chief component of a pasture degraded by too many years of too many rabbits always selecting the best. The central area of a burrow around the entrance holes was kept bare with traffic, it was only the long deep shoots which ran under pasture. Farmers named the wiregrasses for the now obsolete gauges of the plain wire in their fences, No. 8 and No. 9, according to the thickness of the stems.

The middle layer of soil was reddened with clay, and the fleshy tap roots of Blue Crowfoot speared down into it. When the seeds of this sturdy herb were ripe, I would pluck them out singly, lick the palm of my left hand, then lay the pointed seed with a straight 3-centimetre-long tail in the saliva. Immediately the tail would curve as though it had come to life and it would twist the seed over and over on my palm. So those fallen on the ground would bore down into it when it rained.

Sometimes a thick root from a Bimble Box perhaps 50 metres distant traversed this layer and my father turned the mattock over and cut it out with the axe head. Where moisture and phosphorus and nitrogen are all in short supply, Australian trees have to extend their roots an astonishing distance to gain sustenance. Many of them learnt to make use of small amounts of rain by shaping themselves like funnels to catch it and run it down their trunks to a network of fibrous roots encircling the base.

The bottom layer of soil, the somewhat inert support of the furious life above, was hard red clay studded with small rough stones. The rabbits did not dig into it; where it began 1.5–2 metres down, the holes stopped sloping downward and ran horizontally along it.

Those burrows also showed me that the earth does not like to lie naked and that often, as a first covering, it throws over itself something stinging or prickly so that nothing dare disturb it again. The quick cover for the dug-out burrows was Scrub Nettles, a metre tall with toothed dark leaves covered with hairs that stung so fiercely bare legs tingled for days after they had brushed them.

I saw horse teams working across that land. The red soil paddock where the rabbits were was surrounded by black soil plains; the first 10 kilometres of the road to the wheat stacks at the railway station was an unformed track across those plains. One wet year no motor lorry could get near our wheat paddocks. My father and the sharefarmer turned the heavy bags of wheat each sunny day so the damp undersides could dry instead of sucking moisture all the way through the grain to germinate it and ruin it. It rained every few days. The bags were rolled and rolled while the stubble under them began to rot and offer less cushioning from the wet ground. An enterprising teamster came into the district with horses and wagons. He pulled our wheat out, making one load every two or three days instead of the four loads a day a truck could make on dry roads, but he moved it all safely. Sticky mud built up on the Clydesdales' hooves until they seemed to be wearing boots, the long hairs on their feathery legs trailed in the mud which glued them into twisted strands like dreadlocks with little mud balls on the end. When they dried, they knocked together and rattled.

One of the sharefarmers, too, used horses for a couple of years. Too distant from any country school to attend, too young to begin lessons with Blackfriars Correspondence School which did not enrol pupils until they were 7 years old, I used to stand in the box of the header for hours at harvest time while the wheat streamed over my bare legs. As the box filled I stood on top of the rising wheat and my view over the paddock grew wider.

These horses were half-bred Clydesdales and rather light and lively. They had to maintain a steady 3 miles an hour (just under 5 kilometres) because the header was ground-driven by a spur wheel running inside the big main wheel. A header cuts off the heads of crops with a sliding knife and feeds them into a spinning drum which threshes them against grooved plates, then cleans the grain by carrying the straw out over the back of the machine on long, stepped devices known as walkers, and by blowing the light husks away with an air blast directed at the grain and husks on vibrating sieves. Elevators lift the clean grain into a box from where, in those days, it had to be emptied into bags.

These headers were finicky about the speed at which they were pulled: half a mile an hour too slow meant dust and straw in the box with the grains of wheat; half a mile an hour too fast meant cracked grain in the box or grain wasted as flour blown out the back of the machine.

The sharefarmer was always trying to slow down his horses. They were badly trained and, because they were light, he used eight abreast to pull the header when six could have handled it comfortably. When the grain began to crack I would take up a handful and stretch it out to him sitting on an unsprung iron seat below me. He would glance at the handful, curse, and haul back on the reins. "Whoah! Whoah! Steady!"

The reins to horses so harnessed to machinery are not run individually to each horse as they are in coach teams. A Cobb & Co. driver held six pairs of reins in one hand when he had tandem threes to haul him over a mountain. The sharefarmer held one pair of reins only, one of them attached to the nearside ring in the bit of the nearside horse, the other attached to the offside ring in the bit of the offside horse. One day as he tried to steady them the near group of four slowed, while the other group surged forward and broke the chain connecting the fourth and fifth horses. Bewildered and out of control, the two groups of four split; one circled to the right, the other to the left. It was no use pulling on the reins — that would merely have increased the turns. There was risk that the strange pull would distort the frame of the header, even pull it apart. The sharefarmer jumped off his seat and moved as quickly and quietly as he could to the offside horse, took up the rein and held him, so that the other three had to halt too. I was slower out of the box and scrambled off the header on the left-hand side, keeping hands and feet out of meshing cogs. I took up the rein that was dragging along the ground, moved in front of the nearside horse with it, tightened it a little, looked up towards his ears to make sure he heard me, and called as calmly as I could, "Whoah!". He too halted and the three harnessed to him settled down alongside him. The sharefarmer joined the chain with a connecting link and once more we spiralled the paddock, adding to the line of filled bags.

School and war intervened for ten years. I came back briefly from Papua New Guinea to my father's farm, but I could not test the soil with new crops, with new methods of working: my father worked it as he knew it. I gained a Closer Settlement Lease, 904 acres (366 hectares) of black self-mulching soil leading down to chocolate brown flats on the Namoi River. It was a reward for war service, an inconceivable bonus now. Australia then needed a greater yield from its land. These farms were awarded by ballot to those with the experience to work them, so the majority of ex-servicemen were excluded. The farms were not a gift, they were leases in perpetuity with low annual rentals so that the government slowly got its money back. We found the money for fencing, for stock, for buildings, or else we borrowed it from the government at low interest.

There were no fences, no buildings on my block. It was a farm plotted on a map as the first white settlers in Australia took up their land years ahead of the surveyors. My land was soon surveyed: short pegs defined where I must run boundary fences.

For the rest I could lay out the farm as I wanted it with internal fencing, I could build a hayshed, silos, machinery shed, house, fowl yard wherever I thought best. No one had ever lived there, it was a part of big paddocks which had grown wheat or run sheep. There were no imported domestic curses waiting for me, no fleas for the sheep dogs, no snails or slugs for the garden, no ticks for the fowls.

It is a wondrous thing to do to create a farm. One works in consultation with the land. I did not know it then, but the Chinese have had that idea for thousands of years. *Dì lǐ*, they call it, "land management" or *fēng shui*, "wind and water". Above ground, below ground, the land has passages for dragons and white tigers. All must be arranged — or more definitely, not disarranged — so that their movements are never disrupted. They must be able to travel freely and quietly, and there must be places where they can rest to build up *qì*, their breath, their energy, which benefits the whole terrain.

I did not design the farm so mystically. It sloped gently down from a hill on a neighbour's farm where there were wallaroos, not dragons, but I did have to see how water flowed after heavy rain before I made fences or permanent vehicle tracks which can turn the flow of water on easily depressed black soil and wash out new gullies. I recorded the heights of river rises at Gunnedah and Boggabri exactly so that I knew how the river behaved at each height of water. The sloping country flattened out for several hundred metres then stepped down to the river in three wide shelves, each of them cut by a gully or two which took water from a higher point of the river and carried it to a point downstream, filling shallow lagoons on the way. To avoid getting stock trapped between the gullies and the river to be drowned if it continued rising, and also to avoid moving stock unnecessarily right off the river country during low floods, I fenced along the top of the shelves so that I need move them only from those paddocks which would flood. To allow for those high floods which covered all the river country, I hinged the fences so that I could drop them flat and let the water flow harmlessly over them.

So I gradually learnt the river, and I learnt to farm my own heavy black soil with new tillage implements and a new red tractor with rubber tyres and lights and a very hard seat and no canopy so that I frosted on winter nights and baked on summer days. The farming of self-mulching black soil was then so new that Jim O'Reilly, the district agronomist at Gunnedah, was still travelling to farmers' meetings with a piece of string in a bottle of marbles and another piece in a bottle of sand. He explained that if the marbles were particles of soil it did not matter how rich they were, the piece of string, the roots, could not make sufficient contact to draw nourishment from them. Black soil was structured like the marbles. It had to be worked shallow with tined implements to make a firm bed for roots to run in.

Black soil had resisted the high, flaggy varieties of wheat then grown on red soils. The only Australian soil with ample phosphorus, it grew these wheats 1.5 metres high with so much leaf that they needed more water than the soil could supply in the flowering days of late spring and early summer. Crops hayed off and set no grain. But in 1946, Dr W.L. Waterhouse, a plant-breeding genius, had produced Gabo, a high-yielding, fast-growing, short-growing, early-ripening wheat so fitted for black soils that it seemed to be designed, not bred. I grew Gabo as my first crop in 1948. It made farming an adventure. My own soil was responding to new ideas.

Black soil is called self-mulching because its grows loose and fluffy as it dries. The top 5 centimetres can be raked through the fingers. During a drought it yields up so much moisture that the soil shrinks and the surface crazes with cracks 2–15 centimetres wide and up to 5 metres deep. When working on machinery in the paddock in such conditions, we spread a tarpaulin to catch any dropped tools, otherwise a spanner might bury itself for ever. Several times I rescued young lambs from these crevasses. An agitated ewe running backwards and forwards and a bleating from underground marked the spot so I went home to fashion a hook on the end of a light pole so that I could work a rope under and around the wedged lamb and haul it out.

Of the trees, only Myalls, a grey-leaved weeping acacia, can grow on these soils. They do not mind if their roots are torn as the ground cracks, and some of their roots are so tough they resist the contraction and the soil slides around them.

When rain falls on dry black soil, each damp particle swells and puffs up. As it rises, dry soil trickles from it, then it too rises as a raindrop hits it. The whole surface of the paddock stirs about in welcome. It takes a lot of rain to swell the subsoil and fill the cracks. I once walked through half a metre of floodwater that had covered one of our river paddocks for three days. I heard a loud gurgling noise and walked across to see what it was. Water still swirled down a crack so quickly that the hollow cone of the whirlpool was about 20 centimetres across at the top. How deep and how far was all that water going? I felt uncomfortable. It seemed to be hinting of a cave somewhere far underground which needed filling.

I learnt to wonder at the river, I learnt to make good use of it and I learnt to fear it. I did not learn it all on my own. Joan Stephenson came to join me as my wife, a marvellous girl who wanted to live life as exuberantly as I did. She was a pharmacist who intended to remain one so that she did not lose her identity. So there were four eyes to watch the river and in a few years there were six, eight, ten with Kim, Kerry and Mitchell. I cannot conceive of a life without children. It would deny all I know to accept that the most wondrous sowing of all should yield nothing. Joan bore her children as naturally as she expected, with quick deep pain, with joy, with satisfaction, and she had lavish breasts to feed them.

The river at mean summer level was lavish with life. Couch grass grew on the banks and ran down into the water, several water weeds sprawled out of the water up the bank to meet it, so that in some places it was hard to see where the river began. For a metre out from the bank there was a fringe of Red Azolla, light green to deep reddish brown with its dense fine roots hanging down into the water, a cover for shrimps, young yabbies and the fingerlings of Murray Cod, Yellowbelly, Catfish and Silver Perch. Good big fish would bite at these times if one knew exactly where to place the bait. A fat well-fed Yellowbelly would not bother to move to a hook full of wriggling worms if it was placed even 30 centimetres from its mouth. One deduced where their mouths were by the fact that at that one little spot a fish would usually bite. There is always competition for good lairs in a river. If one fish is taken, another moves in to claim its site.

After a shower of rain, turtles would come out of the river to lay eggs on the bank. They would scratch out little tunnels 30 centimetres long and 30 centimetres deep, lay ten eggs or so with measured urgency, cover them up so that no one could see that the ground had been disturbed, then slip back into the water.

A summer flood which filled a dry lagoon or two startled them into life with extraordinary suddeness. Within hours of their filling, frogs and yabbies had come up from suspended life in damp chambers metres down, dragonflies flew in mating loops over the surface, green or red mayflies patrolled the edges, and water beetles by the thousand swam hard to accomplish their lives before the water dried up again.

We began to irrigate land from the river after a trip to a school at Yanco to learn how much water different plants needed, how much green feed sheep and cattle ate a day, and how to calculate the economic adjustments between engine power and spray pressure and labour to move pipes. We grew clovers and grasses which our land had never grown before and we bought cattle and sheep to fatten on them, rationing them to all they could eat in four days with electric fences. Every four days they had fresh lush pasture to move on to, every four hours night and day for ten years I moved irrigation pipes. It was profitable. Very few farmers were irrigating in that area then and I had little competition in buying the stock to fatten. I bought them at a price per pound of their estimated live weight. Since I knew the approximate selling price, how long it would take to fatten them and how much weight they would put on each day, I had to know exactly what I was paying for their initial weight. It was something I could not talk about — I would have been regarded as too outlandish to deal with. Farmers then, and fat stock buyers, made a rough estimate of quality to arrive at a price. The fat stock buyers had a haphazard idea of weights; few farmers had much idea at all. Nowadays almost all stock are sold at cents per kilogram, then they are driven over a weighbridge after the sale.

We grew crops, too, for summer and winter harvest: wheat, barley, oats, grain sorghum, sudan grass, linseed, cowpeas. Forty acres of malting barley, 16 hectares only, bought our first car the year we married — a Peugeot 203, which was an exhilarating vehicle to handle at speeds we had never considered on rough gravel roads in the staid, clumsy and erratic English, American and Australian cars of that time. It is a measure of the change in the circumstance of farming over forty years that it would now take 400 acres of the same crop (160 hectares) to buy an equivalent car.

After twenty years, for one urgent hour the river showed us what I had waited all that time to see. A particularly muddy rise — the water was of the consistency of light cream — forced all the animals in the river to come to the surface to breathe. Shrimps in millions formed what looked like a crust along the water's edge, yabbies poked up through them, and behind them, fin to fin, the fish lined up with gaping mouths, fingerlings 1 centimetre long, Murray Cod 1 metre long. Turtles swam through them and came out on to the banks, so did water rats which stood quietly dripping and watching us as we walked past. Water is so natural to them that they do not shrug it off like a wet dog.

The water cleared a little, all the creatures knew they could filter oxygen out of it again, and they disappeared suddenly under water. I had to go and move stock off the river and drop the fences. The river was beginning to worry us. Once it had come down in a wild flood a metre higher than any records allowed. It tore expensive new netting off fences I thought well clear of flood and rolled it up in a long useless rope, it came into the house and deposited silt studded with small dead fish, it stripped topsoil from one part of cultivation paddocks and deposited it in another so that I had to work soft soil and hard soil in separate blocks with different implements. It took years before the paddocks settled down to an even texture.

Farming was our sustenance, our measure of the world; writing was my reason for living. *Sheaf Tosser*, my first book of poetry, was published and the reviews were a delight to read; *They All Ran Wild*, a difficult, fascinating job which I fitted into quieter jobs on the farm, was about to be published. I had signed a contract with Douglas Stewart at Angus & Robertson to write a human history of the Chinese in Australia. Many farmers on the river had begun to irrigate and the constant work was no longer as profitable as it had been. Research on the Chinese would take me away from home for increasing lengths of time and the river seemed to be threatening the book: "While you're away I'll drown the farm".

We sold *Keewaydin*, the river farm, to concentrate on research, and immediately we felt that we had abandoned life, not a farm. Money in the bank seemed too insubstantial a thing to support years of work. It produced no wonder, no anxiety, no extra knowledge. We saw an advertisement for *Cumberdeen*, a farm at Baradine. We inspected it: it was on red soil far from a river, it carried a fine crop of wheat, it was studded with Kurrajongs, profuse and shapely trees outstandingly deep green against the bleached grey of eucalypts. We bought it. I learnt to work red soil. We engaged an excellent farm worker so that we could all go away on research. There was work to do in every state library in Australia.

Red soil produces crops which are not as luxuriant but are more dependable than those grown on black soils. Red soil yields up more of its moisture to plants trying to set heads in a dry spring, it receives rain more deeply than black soil. The surface does not move about with the same ecstatic reception of rain, it lies still and absorbs water slowly and deeply. I was astonished at how little rain was needed to sow a crop. Twelve millimetres was ample to soak down to the seed bed at 5 centimetres and below it to connect with subsoil moisture. That amount on black soil wet the top 3 centimetres and left a hopeless dry band under it.

Those crops like wheat and barley which developed from grasses could produce their permanent roots on the sowing rain, a great advantage. The first roots grow from the seed which sends up a shoot through the ground. At about six weeks, these roots begin to die and permanent roots develop on the stem about a centimetre under the ground. On black soil the moisture never lasted long enough for these permanent roots to develop, which meant that there had to be another fall of rain within two months or plants died. Red soil crops could survive at least three months without rain.

We added sunflowers, lupins, safflower to the list of crops we grew at Boggabri. We always had paddocks ready for sowing so that we could take advantage of whatever rain fell. One good year we had a different crop to strip in each of seven months.

Buying new machinery was an especial delight. We discarded the small, the worn, the inefficient for a promise of wider cut, of less maintenance, of better work generally. The first disc plough I used required the disc bearings to be greased every two hours which meant every 5 acres (2 hectares) — I used grease by the 5-gallon pailful (20 litres) and bought them frequently. The last plough I owned needed greasing every 400 hectares, every 120 hours. Clumsy levers that required all my weight to adjust the working depth of a machine were replaced by hydraulic cylinders activated from the tractor, a flick of a lever lifted a machine out of the ground or forced it in to full depth and any degree between. Air-conditioned cabs with padded sprung seats saved the tractor driver from jolting, from cold, heat, dust, from noise, from exhaust fumes.

For the first season of such a tractor, one circled the paddocks in wonder at owning it. We never had the spare money to replace machinery as a matter of course: research, travel, school fees, food and wine, the belief that life was to be embraced not suffered did not allow for a daily balance of accounts. We bought machinery when we had to, though we never kept a machine which was likely to break down in the middle of work. In forty years I bought six tractors.

After some years of growing crops only, we began to fatten cattle, travelling to Coonamble, Dubbo, Narrabri, Moree, Goondiwindi to buy them. I calculated the feed in the paddocks with a square yard frame which I had used on the irrigation block (and which I never converted to metrics) to define the areas to be cut and weighed on the set of pharmacy scales Joan used in the kitchen. I bought only the exact number of cattle to match the feed in the paddocks so that no drought could upset their fattening. We were amazed to find that the drier forage of Baradine fattened cattle two or three months quicker than the lush, irrigated pasture. But the first few mobs chewed old bones, the remnants of long-dead sheep that had belonged to a former owner. It showed they were looking for phosphorus so we increased the fertiliser on the forage crops and the cattle fattened quicker than ever. It is satisfying to buy wild cattle so poor that they look long-legged and narrow (herring-gutted), then to send them to market as quiet, wide, short-legged beasts. To quieten them (quiet is an essential state of growing fat), I would first drive around them two or three times a day until they got used to the vehicle, then I would stop and open a door, talk to them, perhaps get half out. After three weeks they would let me walk round them, always talking to them, always moving slowly and quietly.

I wrote *The River* as a farewell to *Keewaydin*; I wrote *Running Wild* for children and *The Green Mosaic* in memory of Papua New Guinea. I wrote *Miss Strawberry Verses* for children and *A Million Wild Acres* to explain the great forest on our northern boundary. In a hundred years abandoned sheep runs and cattle runs developed a primeval state — what is young seems ageless. In a hundred years this farm could eradicate me and those who worked it before me. Is the soil angry with us, anxious to be rid of us? I do not think so. It merely demands that we be attentive. Those who failed over the great expanse of the Pilliga Forests did not understand it. What they did was to define the area of good land. Our northern boundary sharply divides acid country from fertile neutral.

I continued with research on the Chinese. It became a vast subject: the more work I did the more there was to do. I had to know enough to create a people and their associations. Sometimes I went away on my own, sometimes Joan came with me and worked as a relieving pharmacist. The children joined us in school holidays. I wrote *Celebration of the Senses* for the joy of living, then Joan died and the durable solid world became shapeless and transitory. I continued work, I had to, or else I would negate Joan's life as well as my own.

Elaine van Kempen came to sort the great mass of Chinese papers so that I could begin writing. They were in suitcases, in cardboard boxes, in piles on floors or on top of cupboards with rough but accurate manuscript notes of what was in every pile, except for one big suitcase marked "NOT READ YET."

I began to write *Sojourners*, the story of China and Australia. We worked well together, Elaine and I. She stayed on when the papers were sorted into chapters to come on a last long research trip to the site of Chinese mining in Queensland and the Northern Territory, we fell in love, we married. The world solidified again. I did not marry a substitute Joan: save for her humanity, Elaine is different in every way. Our bed, our food, our work together and apart are new experience. I wrote *Doorways* to register an extraordinary year together; at the Warana Festival, Brisbane, in September 1990 I told this poem for her. It was one week old, not even Elaine knew of it.

The tall woman with the kind face
Which could not shape ungentleness,
The kind woman with the long black hair,
I met her when I needed her;
We worked and loved together.
And the old house which had begun to droop
Without any laughter to hold it up
Restored the fabric of its walls
With new talk, new skills, new cooking smells.

Cumberdeen must now be sold to write *The Growth of Australia: the shaping of a country*. I have a Creative Fellowship and a publisher's contract to write it which are no more than background to a compulsion to write it. It will keep us travelling all over Australia for three years, and a weatherboard house with irreplaceable books and papers and a few fine pieces of Elaine's furniture cannot be left to care for itself.

It is a bitter thing to sell one's sustenance of more than twenty years. But disruption is part of a writer's instigation. To grow comfortable is to die. Without the move to Baradine I would not have gained *A Million Wild Acres*, perhaps not even *The River,* since the story would have had continuance, not completion. So we move to Sydney as a depot, to the indignity, the inconvenience, the restlessness, the compression of city life. I have to abandon what I have loved. In the years ahead of writing *The Growth of Australia*, the book will be my purpose, my retreat.

As Joan and I left *Keewaydin*, our indicator of spring, the willow tree I planted at the end of the kitchen drain suddenly died and blew down. As Elaine and I prepare to leave *Cumberdeen*, an especial Kurrajong is about to fall over. When I came here it was an upright tree growing out of the top of an old box stump about 5 metres high. Probably a Currawong had dropped a seed on top of the mud guts of termites and the soil which blew in to fill the stump. We could see its roots through cracks. The tap root was a core about which the constrained surface roots coiled. Year by year as the roots swelled, strips of the stump cracked off and fell. Now there is no part of the stump standing. It lies on the ground and the tree leans more heavily over it every day. Its trunk is the plaited cable of roots which are not strong enough to support it. In a few months the tree will collapse. Longicorn beetles and jewel beetles will lay eggs in it and their big grubs will bore through it, helping it rot so that it can restore to the soil what it took from it. The cycles are inexorable.

The song of the dingo

Aborigines living near the mouth of the great Victoria River in the Northern Territory used to dance the arrival of dingoes. One saw them running excitedly up and down the decks, stopping to look towards the land, down at the water. They jumped overboard, paddled ashore, shook themselves dry, a couple rolled in the sand, then they all began to nose about and hunt. By the light of the fire the players seemed to become dogs. Those who watched this marvellous re-enactment always presumed that it told the story of how Aboriginal ancestors brought the dingo to Australia. Indeed, although there is no evidence whatever that a reinforcement of Aborigines arrived about 4000 years ago (which fossilised bones suggest was the time of arrival of the dingo), it was generally presumed until very recent times that Aborigines brought the dingo. But the story that the dancers were telling was of the arrival of the dingo with visitors, not black settlers.

In a fascinating study of dingoes which extended to dogs and wolves throughout the world, Dr Laurie Corbett, now stationed at Darwin with the CSIRO, made the startling discovery that these visitors took their dogs back with them, since he found Asian dogs infested with *Heterodoxus spiniger*, a species of sucking louse which evolved on kangaroos.

The idea of a long association of South-East Asia with Australia still discomforts many historians, although it is so likely that it would have been extraordinary if it had not happened. The Aboriginal stories of Baiini, a small, yellow-skinned people who made regular visits, were relegated to Dreamtime myths instead of being accepted as oral history.

The regular visits by Macassans to collect and prepare trepang for the Chinese market did not begin until about 1670, but carbon dug from under the mangrove charcoal where they boiled their huge cooking pots suggests that somebody else, possibly Chinese, worked there in 1200 A.D. and again about 1450. Those Indonesian fishermen who are now being towed in their ships to Broome and Darwin harbours, where they are gaoled for illegal fishing in Australian waters, are following routes which their ancestors explored thousands of years ago. Although this modern fishing is mostly engineered by ruthless and wealthy businessmen, the poor men they send out in insufficient craft believe they have rights by long usage. Since men from Indonesia first began venturing to sea, which is as long as there have been men in Indonesia, they have regarded Torres Strait, the Arafura Sea and the Indian Ocean off north-west Western Australia as home waters.

Some of these many visitors brought cats with them; there were wild cats in northern Australia and in numbers along the central coast of Western Australia long before white settlement. Apart from the fact that the Aborigines of those areas regard them as native, when working on *They All Ran Wild*, the story of Australia's introduced animals, I found several mentions of considerable numbers of cat tracks in parts of Western Australia where it was much too early for them to have spread from settlement. Those cats might have come from Dutch shipwrecks, perhaps even from Phoenician wrecks.

But who came and returned with dingoes 4000 years ago is not known. The lice indicate that they were from South-East Asia; and besides, it is the practice only of people from this area and from the Pacific islands to carry dogs as usual passengers. They make practical pets, pleasant to have on board, good food when needed. It was, it is, the custom to set out from port on long voyages with two or three bitches in pup.

The dingo came to Australia about 2000 years after it evolved. Its ancestor is the palefooted Indian Wolf which several different people domesticated as they settled down from wandering hunter-gatherers to fixed-place farmers. Some races bred them selectively for the traits they wanted — the ability to herd cattle, for instance, or the quietness which makes a good pet or the controlled fierceness which makes a good guard — and thus they changed them into distinct breeds. The people of South-East Asia mostly left them alone; only circumstances evolved the dingo.

The features which distinguish them are mostly in the flesh, not the skeleton which, apart from consistent differences in the size of the skull, is the same as that of other dogs. They mate once a year like wolves, not twice like dogs; they cannot lay down their short, stiff, triangular ears but they can turn them in the pricked position; they cannot bark, managing sometimes a few whining yelps — the chief call is a long-drawn howl, higher-pitched and more dismal than that of any domestic dog. Australia produced four remarkable animal noises: the seldom-heard scream of the Barking Owl so much like a woman at a high point of terror; the raucous clatter of the kookaburra; the high, thin, reedy, mournful cry of the Stone-curlew (now ignominiously tagged the Bush Thick-knee); and the long-carrying howl of the dingo which can lift the hairs on the back of one's neck even in the daytime. It used to be one of the commonest of bush voices. Dingoes thrived in Australia; the descendants of the northern introductions soon occupied the whole country. They are an impressive, athletic animal. The head is blocky and the jaws powerful; almond eyes are keen. The tail is a short bottle-brush, a strip of coarse hair runs down the back with an undercoat of fur as wide as a hand. Fine soft hairs stand out from the sides. They made rapid changes to coat and colour to suit different environments but these are temporary expediences, not the permanent changes which indicate sub-species. In snow country coats grow long, dense and a rich reddish-gold; the sparse coats of those in the hot Centre are yellowish-fawn to match the land. Some dingoes are pure white (and not albino), some are black with tan points. Most dingoes have splashes of white somewhere, usually on the feet and the neat point of the tail.

Expert hunters with omnivorous appetites, they must have had a startling effect on Australian animals, especially the kangaroo family which had previously been preyed on only by Aborigines, by Wedge-tailed Eagles and by the slow, long-winded Thylacines which hunted alone and were never plentiful. These strange creatures, looking and moving like a cross between a dog and a kangaroo, could not meet the competition and soon disappeared from the mainland. Undoubtedly, the intelligent and cunning dingoes not only lessened the available prey, they would have patiently trotted after a hunting Thylacine then stolen its kill, mauling the Thylacine if it showed fight.

It is also likely that the mange dingoes brought with them attacked Thylacines. With no previous exposure to the debilitating mites, they would have suffered cruelly. Dingoes harbour two types of mange: sarcoptic, which causes coarse, weeping lesions on the skin, loss of hair and great itching; and the even more serious demodectic mange caused by mites, which complete their life cycles within an animal's body, thus keeping up an awful destructive circling inside and out.

Aborigines naturally adopted dingoes as hunting aids, as companions, as warmth on freezing desert nights. There soon began the long, strange, inadvertent association of Aborigines and dogs. They treasured them, they treasure them, but they show no rational concern for their welfare. A man would carry an injured dog 20 kilometres back to camp, then give it nothing to eat except a bone which he had first gnawed clean. A woman would feed an abandoned pup at her breast, then leave it to fossick or fight for whatever else it needed. Old men and old women welcomed them as sleeping companions — the more dingoes they had about them the warmer they slept — so they broke their forelegs to keep them in camp. When they moved, they might carry one or two and leave the others to die. Each spring there were always pups to be found in the bush to replenish the supply.

Wild dingoes soon learnt that there was good scavenging around Aboriginal camps: bones, the skins of reptiles, the cases of cooked yams, human excreta. Keeping well out of sight, they followed Aborigines on the move. When Europeans arrived, they raided their camps, often making themselves destructive nuisances, even eating stirrup leathers and tearing the sweaty lining out of saddles. In 1966 when we went into the Centre on research for *They All Ran Wild*, I rang Eric Sammon, then in charge of Birdsville police station. "Don't come out here without a gun," he told me. "And don't let dingoes into your camp. They'll eat anything, even your bloody kid if you've got one."

It is unlikely that the dingoes attached to Aboriginal camps bred there. Apart from the fact that they were usually too poor and too mangy to breed, conditions were not private enough. Mating and care of pups are remarkable rituals. When a bitch begins to come on heat in late autumn, she soon collects a following of dogs, up to a dozen, who can smell her secretions from an astonishing distance. Furthermore, she advertises her condition by special howls and strategic urination. For a week or more the bitch keeps the males at a distance, but they all hunt together, usually with the bitch as leader. Then she accepts one mate only who stays with her to rear the pups. Mating is protracted as in the domestic dog, the organs are similar. The long thin penis is supported by a bone, two bulbs at its base swell out into a dumb-bell-shaped lock which fits into corresponding sockets in the vagina and holds dog and bitch clamped together for about twenty minutes. During the mating the semen changes remarkably in consistency, quantity and composition. All the spermatozoa are expelled in a thick slime after a few vigorous thrusts in the first eighty seconds, thereafter several ejaculations provide about 30 millilitres of a thin liquid which gives the millions of spermatozoa a happy medium to swim in towards their goals.

Even when living in the loneliest country, bitches retire to have their pups. They seek out caves, hollow logs, clefts in rocks or the cover of dense lignum along rivers and flood channels. They never dig breeding stops like foxes, but they frequently enlarge rabbit burrows in sandhills. The number of pups varies from one to eight, there are usually three or four. The bitch stays with them for a couple of days only, then she moves several hundred metres away and camps with the dog, always downwind so that she can hear and smell any intruder near the pups. She does not risk taking them to water until they are a couple of months old and able to run freely. When her milk supply begins to wane after about four weeks, she carries food and water to them. Her mate helps her make a kill, she runs to water, has a big drink, comes back to the kill, eats until her belly distends, then she trots back to the pups and regurgitates it for them. The dog seldom helps with this feeding but once the pups are big enough to be taken to water and to tear their own meat into swallowable chunks, he helps the bitch carry flesh to them from distant kills. They deliberately conserve the close game for the pups. When the pups are about eight months old, dog and bitch together teach them to hunt. Already they would have learnt to find fruit, rats and small lizards but considerable skill is needed in hunting adult kangaroos. Dingoes mostly take joeys on their first bewildered essays from the pouch but, when hunting in packs or even pairs, they seek bigger game. Kangaroos are of equal speed and equal stamina, so a bitch will take a kangaroo on a long run, then turn it at an angle to meet the dog who has trotted across the diagonal. The dog then takes his turn, runs the kangaroo hard and turns it back towards the bitch who has caught her wind again trotting across another diagonal. It is when the kangaroo finally bails up that there is danger. Even big dogs like the early nebulously bred Staghounds often met disaster in the arms of kangaroos. They attacked from the front, leaping for the throat, and the kangaroo, bracing itself, threw its powerful arms about the dog and crushed it. If it were near water, it carried the dog into it and held it under to drown.

Dingoes are much too experienced to be so treated. One goes in front of the kangaroo to attract attention, snarling and challenging safely out of reach, while the other leaps in, tears a piece out of the butt of the kangaroo's tail and leaps back again. After repeated bites the tail dangles uselessly and the kangaroo cannot rear to defend itself. It falls helpless and the dingoes rip at belly and throat.

When teaching pups, a pair of dingoes choose a kangaroo of a size that the pups can handle, then they partly disable it and drive it back for the pups to practise on. The behaviour of sheep took all the hard work out of teaching pups to kill. Instead of separating like kangaroos and hopping in different directions when danger threatens, sheep mill together in tight flocks. A dog and a bitch can make one rush, tear the flanks out of several ewes, then encourage the pups on an exciting chase after entrails dragging through the grass.

Dingoes decided the flock management of the first white settlers. Neither cattle nor sheep could be left unattended night or day. Naturally, when there were no fences, cattle had to be tailed and sheep shepherded during the day. But any good stockman woke frequently at night to check that the calves of the cows he had camped on water were not being worried by dingoes; a shepherd maintained vigil with a rifle beside him in his watchbox, a portable covered bunk. A flock was his charge, he had to pay for lost sheep, and no squatter was concerned about how much sleep he got.

It is not in the nature of a dingo to kill more than it can eat. Since putrid flesh is acceptable to them, they will take an opportunity of securing a couple of weeks' feed. But the different behaviour of sheep, the strange circumstance of sheep in yards, confused and excited dingoes. They killed unthinkingly. A pair of dingoes in a yard might well account for a hundred sheep before they tire of the sport.

Lambs, so easy to catch, so meaty, became an especial delicacy. Flock owners set out to eradicate dingoes by shooting, hunting them with dogs, poisoning, trapping. They were singularly successful; by the early 1860s there were few dingoes left in any of the closely settled districts of southern Australia. Kangaroos and wallabies, favoured by the new waters established for sheep and cattle and by vast areas of grass trimmed by sheep to the length best for kangaroos (who had previously been restricted to areas burnt for them by Aborigines), took full advantage of the lifting of the predation of Aborigines and dingoes, and increased enormously. Men who only a few years before had saved small flocks from the skin-getters at the instigation of their wives, found themselves running more kangaroos than sheep. In organised drives they yarded and slaughtered thousands.

When sheep extended on to huge runs in western New South Wales, western Queensland and the north of South Australia, dingoes ravaged the flocks, reducing lambing percentages to about 30, killing adult sheep by the hundreds. The three governments built a long 2 metre high fence to separate cattle from sheep. North of the fence in cattle country dingoes could be tolerated, south of the fence in the sheep country they were to be eradicated. Boundary riders were stationed along the fence 40 to 50 kilometres apart to maintain it — a constant, difficult job where there were flood plains and shifting sandhills; bounties were placed on the scalps of dingoes, doggers were employed to catch them; Wild Dog Destruction Boards set up to organise the work. Off and on, bounties were placed on dingoes north of the fence. For years Aborigines harvested pups by the hundred. Their scalps became local currency even in dealings with European storekeepers in Alice Springs.

Trapping and shooting dingoes, especially of known and named killers for high bounties, became a way of life, a secret lore, for numbers of men. They had to be expert trackers, they had to be able to call dingoes to them, each of them developed his own lure to attract dingoes to a trap. In the mating season, a sprinkling of the urine of a domestic bitch on heat was often successful, others expressed sebum from the glands about the anus of a domestic dog and either smeared it straight on to a grass clump near the trap or extended it for use near several traps by mixing it with urine. Some made up an unsubtle, nauseous mixture of dogs' turds, sardines and condensed milk and left it to ferment for months.

No matter how carefully the big steel-jawed traps were set with every layer of dirt, every piece of stick and blade of grass put back as it was, many dingoes were always aware that they were there. Some trappers set a trap a little obviously, then set a blind trap with great care where they estimated the dingo would walk when dodging the first. Bill Baldwin, senior dogger in the Western Division of New South Wales when I met him at Tibooburra in 1966, told me of one bitch who outwitted him for months. No matter how carefully he set a trap she always located it — sometimes, when uncertain about exactly where a trap was, she seemed to make a point of finding it, pushing her forepaws carefully through the sand until they touched metal. One morning he noticed that she had used the left-hand track of his four-wheel drive as a pad. Expecting her to go the same way again, he set a trap in the track, carefully replaced the topsoil, then etched the tyre pattern over the trap with a stick. The next morning he had her.

Aborigines soon renounced dingoes as camp companions and adopted European domestic dogs as more friendly, more obedient and less independent. And they certainly bred in the camps; many tribes soon had hundreds of dogs.

There was early European interest in dingoes as pets. John Hunter, who became our second governor, had pups in his care in 1788. In 1830 pet dingoes left off the leash were a nuisance in Sydney streets. They joined with packs of mongrel dogs to menace horsemen and pedestrians. Dingoes do not make good pets. One person can handle them, but they seldom extend their trust to anyone else.

Since 1981 an abandoned dingo has scavenged rubbish bins in the old nightsoil lanes of Sydney's Redfern. He howls at night, bringing back a call that had disappeared from there 150 years ago. He answers to the names of Tiger or Whitey — there is a white blaze down his golden chest — and children pat him briefly but he needs no closer attachments. As a young girl in the 1940s, my wife Elaine went to sleep each night to the howling of dingoes at the Lone Pine Sanctuary at Fig Tree Pocket, an outer suburb of Brisbane. Claude Reid, the owner, had several enclosures of dingoes and he mated some of the bitches to Strongheart, a fine German Shepherd who used to go down to the Brisbane River jetty with a koala on his back to meet visitors coming to the park by launch. They were remarkable matings since there is strong antipathy between German Shepherds and wolves. The great fear of such crosses producing a superior strain of wild dog — it led to a law prohibiting the keeping of unsterilised German Shepherds in country districts — was an absurdity.

The hardworking and intelligent Australian Cattle Dog was bred from blue-speckled Highland Collies crossed with the dingo with later introductions of Dalmatian and Kelpie. The Bagust brothers of Sydney "bred a lot and drowned a lot" but by 1890, fifty years after the first crosses were made, the type bred true.

Dingoes, male and female, do not mate readily with any domestic dogs; however, sufficient mixed breeding has taken place to put purebred dingoes in jeopardy. Only in the deepest parts of forests in New South Wales and Victoria are there now true dingoes; the thousands of dogs on the forest outskirts are dingo crosses and domestic dogs run wild. The necessary poisoning of these in the interests of animals domestic and wild threatens the last dingoes, already threatened by crossbreeding. Laurie Corbett believes that unless something is done urgently to protect the dingo, it has a life of perhaps fifty years, 100 at the most. The country with the purest dingoes is now Thailand, since it has very few domestic dogs. In the first mating of a dingo and a domestic dog, both have to overcome great natural fear; it is much easier for their crossbred pups since they are more acceptable to both dingoes and wild domestic dogs. Interbreeding soon becomes commonplace.

The many dingoes in Kakadu National Park in the Northern Territory are still pure — that is the only part of Australia where one can now be certain of hearing them howl — but there are 500 domestic dogs (as well as many cats) living with Aborigines in the park. In the central deserts, pure dingoes are now being exposed to domestic dogs for the first time. Aborigines in search of better lives are moving out from distressed mixed communities such as Balgo between the Tanami and Great Sandy Deserts (despite its troubles, that community is producing some remarkable paintings) and they are taking their dogs with them. At Balgo recently I stood in one place and counted 120 dogs in the shade of nearby buildings. Perhaps I could see one-tenth of all the dogs. There are more dogs than people in the community, some of them so mangy that they have not a single hair on their bodies. They seem to be covered with thick, wrinkled leather.

After 4000 years the dingo belongs in Australia. It requires protection for its own sake, but much more important than that, it had fitted the scheme of natural order as a major predator and it cannot be removed without disruption. Foxes, cats and wild domestic dogs cannot take its place without bringing insupportable new tensions to an environment changed so much by European settlement.

Some scientists now believe that dingoes keep foxes in check. The fox population is now higher than it has ever been. They thrive during droughts because of the number of dead animals to feed on, the thousands which used to be shot each year for a valuable skin trade are now left to breed because of the stupid proscription against the wearing of fur. Instead of being something of a resource, foxes are now another pest to be controlled; and few landowners can bear more expense.

There is real danger that foxes will soon eradicate several species of small and medium-sized marsupials, including the Tiger Quoll.

Dingoes must be brought back to state forests and to national parks and nature reserves: they belong there. These areas which ought to maintain natural Australia are insufficient and ineffective without them, since kangaroos and rabbits breed out of control. It would be an expensive thing to do, because domestic dogs will have to be got out of the parks, then kept out with double 2 metre netting fences reinforced with electric wires which will also keep the dingoes in. It would take many years, but dingoes can restore an accustomed balance to much of Australia.

Trees trees trees

On 20 July 1989 Bob Hawke stood at Wentworth on the Murray, from where he could almost command three states, and shouted a program for the growing of a billion trees. It offended bodies like the Potter Foundation and the Natural Resources Conservation League and Greening Australia that had been planting trees for years, it ignored the work of J. Ednie Brown who planted still-standing model forests of River Red Gum in South Australia in the 1870s, it overlooked the vision of Wilf de Beuzeville who foresaw a colonnade of trees 1000 kilometres long and 1.5 kilometres wide through central New South Wales in the 1940s, and it made no provision for essential scientific studies of our forests. Yet it was a valuable pronouncement. It mustered a wondrous new enthusiasm for the planting of trees.

The full statement was published as *Our Country Our Future*. It dealt with drift netting, Antarctica, the Greenhouse effect, the ozone layer, land degradation, aid to other countries, housing, disposal of waste, tourism, mining. The writing is atrocious: important declarations disappear in a mist of unnecessary words. The only survivor is the One Billion Trees Program, 400 million trees to be planted as seedlings by the year 2000, 600 million to be sown as seeds or regenerated naturally behind protective fences. Most city newspapers, so many wise heads directing pencil and paper, have calculated the impossibility of it: 11 415 trees set growing every hour for the next ten years.

It is possible to handle extraordinary numbers of trees. Australian Plantations, south of Lismore, is growing Teatree for the oil so much in demand for cosmetics and pharmaceutics. The main species, *Leptospermum linariifolia*, Snow in Summer, grows only 1.5 metres high: it can be planted densely. Already the company has 15 million trees on 1046 hectares; by the end of 1992 they will have 42 million trees on

1700 hectares. The nursery produces 10 million seedlings a year, the planters can handle 150 thousand a day. So it takes one company nine weeks to put in 10 million trees on a small area of the north coast of New South Wales.

Early in April, Elaine and I decided to assess the national tree-planting program. We travelled from Baradine down the Lachlan River to Booligal, then south through Hay and Deniliquin, across the Murray River to Echuca, down into Victoria to Hamilton in the marvellous Western District, then north through the Mallee to Mildura, up the Darling through Menindee to Wilcannia, then east through Cobar, Nyngan and home. Elaine had much information from Gus Sharpe of Greening Australia which is managing the scheme and from Rob Youl, his counterpart in Victoria. We spoke to their field officers, to members of the CSIRO, to agricultural scientists and managers of forests in New South Wales and Victoria, to the distributors of irrigation water, to farmers, graziers, teachers, school children, to nurserymen, sawmillers, vignerons, orchardists, to groups formed to plant trees, to the members of "Rural Trees Australia" who are architects of exciting new farms. When we got home we rang other parts of New South Wales and other states. So much enthusiasm, so much expertise is being brought to growing trees from Cape York to Albany that it is likely the number of one billion will be exceeded. After all, it is an American billion, not an English billion which would be a thousand times more.

We left our own well-treed farm where there are natural green roads and windbreaks that I planted, we drove beside sections of the great Pilliga Forest, through the Goonoo Forest to Dubbo, then through big belts of Myall growing in rich reddish loam on the Narromine–Condobolin road, through clumps and belts of Belar broken by Bimble Box and White Cypress Pine, through country well-dotted with Kurrajong, through Wilgas trimmed underneath by sheep, over poor gum and ironbark hills, and down into more Kurrajongs. That was 350 kilometres of the trip. It seemed that Australia did not need another tree. Indeed there are now many more trees than there were at the time of white settlement. Statements such as "The CSIRO estimates that two-thirds of our forests and bush have been cleared in the last 200 years" give a broad, lazy idea that tells nothing of the fascination of what has happened. The whole country is degraded and disorganised. There are now too many trees where there was open woodland, there is open woodland where there were dense scrubs. This is what has happened in Australia, this is what is happening.

If one brought back one of the early explorers and took him over his old tracks, he would say, "These trees were not here before. That hill over there was all grass. Now it is covered in Bimble Box." But then he would say, "Where are the shrubs? There were patches we cut our way through. I named a lot of them. And the flowers? There were hundreds of acres of Garland Lilies here. They were so beautiful I specially mentioned them. And the grasses are different. Our carts bumped over great clumps of Oat Grass. And where are the native herbs? All I can see is rubbish from England and thistles from Scotland. And what have you done to the ground? It's hard. It sprang under our feet. All the topsoil was loose."

One begins to meet worried people on the outskirts of the Riverina. This remarkable region was the active monitoring system for all the surplus water above ground and underground for an area 1400 kilometres long by 800 kilometres wide from southern Queensland to central Victoria. It cleaned floodwater of silt, it cleaned the land of salt present naturally in the soil. Each year it handled thousands of tonnes of salt. It stored it in lakes and in waterholes in the Lachlan and Darling and other rivers that stopped flowing in late summer, then flushed it out into the Southern Ocean with spring floods of fresh water that it passed through a complex of filters so that they did not take too much soil with them. The workplace between the Murrumbidgee and the Murray is constructed like a delta. The streams wind, often covering three times a straight-line distance, so they command a lot of ground. There are rivers, swamps, lagoons, anabranches, short creeks with individual heads in little hills, creeks that come out of one river and join another. The Edward is an anabranch of the Murray, the Wakool is an anabranch of the Edward. In an area 300 kilometres by 150 kilometres it is difficult to move 20 kilometres in any direction without crossing a stream of some sort.

So well watered, yet so well drained, lying between the 200 millimetre and 400 millimetre isohyets, the country usually looked dry. The explorer Sturt, never a hopeful man, found it desolate. Round-leaf Pigface lay in a deep pink carpet over flat, treeless plains. Sometimes it seemed to be the only plant in hundreds of hectares. Then it gave way to great grey plains of saltbush, *Maireana, Atriplex, Rhagodia, Enchy-*

laena, Malacocera, five genera with many species with names like Old Man Saltbush, Ruby Saltbush, Cottonbush, Bluebush, Soft Horns. They were annual and perennial shrubs to 2 metres high, little plants 10 to 40 centimetres high, a few climbers, prostrate mats. Samphires and glass-worts, similar fleshy plants, grew among them all. There were grasses, too, the showy White Top, *Danthonia caespitosa,* on the plains with some Wallaby Grasses and Panics and *Stipa elegantissima,* Feather Speargrass, all stem and lovely flower, and on low-lying patches the shivery Fairy Grass, *Sporobolus caroli,* and Sand Brome, *Bromus arenarius,* on the long fixed red sandhills that repeatedly break the grey plains.

These sandhills were vigorous areas, the breeding ground, the nursery, the protected home of insects, birds and mammals. White Cypress Pine grew on them, several Hopbushes, Emu Bushes (*Eremophila* spp.), tall Needlewoods (*Hakea* spp.) and the tall Cooba or Native Willow (*Acacia salicina*). Some Grey Box grew on them, too, and Yellow Box wherever the soil was richer. In many places tongues of sandhill plants extended a couple of kilometres on to the plains. Nearly all this growth has gone. The ridges are barren rabbit warrens. I do not think it was deliberately cleared. Rabbits ate it out in the 1890s and there have always been enough to stop it coming back.

Creeks and rivers, lakes and lagoons are delineated by River Red Gums, the magnificent trees that mark out almost every water in Australia, yet the botanist who named the species in the 1830s did not describe it from an Australian specimen but from a single tree grown by the monks of the Camalduli order in the Apennines out of Naples, so it is *Eucalyptus camaldulensis.* During summer floods these trees drop millions of their tiny seeds into the water swirling about their trunks and they germinate under the mulching wrack on flood edges. Sometimes conditions were right for whole plains to spring with seedlings and over forty years it became a forest of tall straight-trunked trees covering up to 40 000 hectares. They grow quite differently under these conditions to the fat, sprawling individuals that fasten river banks. This is how the great complex of Millewa–Barmah Forests began. They spread on both sides of the Murray between Echuca and Tocumwal.

Together with sedges and grasses and forbs that grow under them, these forests are now the main filters for Murray waters. The heavy brown flood that opened gates let into the eastern boundary of the Millewa–Barmah conglomeration of state forests reveals the bottom at a metre depth after it has travelled 40 kilometres to the western boundary. The lost filters, extensive marshes that are now drained, were equally effective. The Lachlan strained its headwaters between present Forbes and Condobolin, spreading out on to wide, shallow flats overgrown with Cumbungi and various rushes and sedges, then it reformed its cleaned water and filled and scoured a number of lakes. Muddy creeks joined it and it made another cleanup through Cumbungi, vast pink beds of polygonum and Giant Rush up to 4 metres high that extended 80 kilometres to the junction with the Murrumbidgee which itself was filtered through kilometres of the same plants extending widely upstream from both banks. The only marshes still performing any of their proper role are those on the Macquarie and they have been grossly interfered with.

Those we first met in the Riverina, members of Trees on Farms Hay, are simply planting windbreaks for the sheep on their bare plains. Without shelter from sudden cold, wet wind, newly shorn sheep can die in hundreds. Brenda Weir showed us Darcoola on the Maude road, a property of 10 000 hectares with an annual rainfall of 250 millimetres. By lowering the stocking rate and spelling paddocks, her husband, Bob, is getting saltbush back on the property. Rabbits and overstocking almost eradicated it in the 1890s. The rigid, blackish mounds of Bluebush dominate, though there is one beautiful paddock of Bladder Saltbush and the grass White Top. The saltbush is still out of order, single species dominate, but everywhere one can find other species coming back, Old Man and Pearly especially. Brenda is planting fenced-off, angled windbreaks with about 80 to 300 trees in each, a mixture of local eucalypts, acacias and shrubs. She plants in August when the heaviest frosts are over on land fallowed deeply for twelve months and she plants them about 10 metres apart so that they do not have to compete for moisture. She shields each seedling with a plastic tube that protects it from hares and rabbits and also collects most of the transpiration and drips it back into the soil. Watered at planting, the trees get two more light waterings in the first summer, then no more. Brenda does not want them to grow for twenty years and then die in a drought. They have to send their early roots deep in search of moisture.

Black Box, *Eucalyptus largiflorens*, that she planted in 1986 is now 1.5 metres high and very healthy.

The land she is working on has probably never grown trees before. The beautiful curving windbreaks will astonish the soil. But it would probably be better to work on easier ground, the sand ridge that runs the length of the property. There are remnants on it of what was there. Mere fencing off would bring many of them back but first the rabbits would have to be got rid of. Already it will cost a year's wool clip to rip the warrens. If they get any worse it will cost many clips. Then there would be 20 kilometres of fence to run but solar-powered electric fences are not too costly and it could be done by degrees. This group has only begun its planting. I am confident the ridges will be restored. The big, quiet plains of the Riverina will get their busy divisions back.

Jan and Brough Gibson, members of the same group, are planting for a different reason. They live right on the Murrumbidgee and, together with their neighbours, they grubbed out trees for more convenient irrigation, so with wide, fence-line plantings they are making restoration to the land of far more than they took. Brough's sister, Elizabeth, does not like the ordered rows of new trees. She prefers natural-looking clumps. Rows will disorder themselves in time with self-sown seedlings and scattered seeding-down on specially prepared land is likely to replace much planting anyway. Yes, clumps look better than rows. I do not like too much order either.

Mike Thompson and Arthur Murphy, foresters at Deniliquin, are experimenting with River Red Gum. They have ground ready in the 40 000 hectare Millewa complex to plant 120 000 seedlings that will be irrigated with differing amounts of water to find out how much those trees need to make good growth. They are valuable timber. They supplied millions of sleepers, they built bridges and wharves since they last particularly well in water, they built barges with finely pointed prows that paddle steamers towed to riverbank sawmills with newly cut logs slung in chains in massive unwieldy bundles over each side, their dead wood powered boilers and now has a big sale as firewood, they made excellent charcoal, and chips from waste timber make salable garden mulch. It is difficult timber to handle, not seasoning easily. Logs crack in the manner of the web of an orb-weaving spider with concentric circles joined to radiuses. Yet the timber polishes beautifully and it makes spectacular thick counter tops. The cracks add character to it. Lately it has been found that freshly cut logs can be peeled before they crack into rotary veneer that can be seasoned overnight.

Not even this great forest can now act with its proper influence as a filter. It seldom gets a flood, it seldom even gets enough water. The foresters have to make do with rain rejection, what irrigators order when it is a week upstream in the Murray then do not want when rain intervenes before it gets to them. So the forest is changing. Silver Wattle, *Acacia dealbata*, is spreading into it from higher dry ground, and also the extraordinary Dwarf Cherry, a root parasite that grows into a bright green spreading shrub and bears round hard, fruit that looks like a seed on the end of a wide, succulent fleshy stalk that looks like fruit. It takes over as a dense understorey in places and weakens the trees it feeds on.

Patches of gums are already weakened by too little water. River Red Gums need floods, preferably water 1–2 metres deep every one to two years. They can take summer-to-summer inundation, but not from the beginning of one summer to the end of the next. Lake Mulwala on the Murray surrounds thousands of dead gums. A weir completed in 1939 flooded one of the filters, a 6000 hectare forest. Neither can the gums live if floods are more than ten years apart. Too little water kills as readily as too much.

Drier conditions lead to another danger: fire. Southern River Red Gums produce no lignotubers, a eucalypt's final defence. The Aborigines exploited this susceptibility. They welcomed some forest, they realised its importance as a breeding ground for the ducks they ate, but they did not want it to extend. They burnt to exterminate flood fringe seedlings. For twenty years I lived on a river block on the Namoi River. Almost every flood produced a wide line of River Red Gum seedlings. I had to plough them out or in no time I would have had a forest, not a farm.

They are very different places, these flooded forests, to dryland forests. There are no ground-dwelling termites or fungi to break down forest litter. Water creatures work on it, turning it into food for fingerlings and ducklings while they themselves supply whatever creatures are bigger than they are. Three thousand breeding pairs of ibis nest in the Millewa complex, 100 pairs of the Royal Spoonbill. Arthur Murphy oversees the nesting and adjusts the level of the water about the ibis platforms so that it does not rise and drown the chicks or unsettle them by receding too far. Cormorants nest in saplings on the edges of the swamps and lagoons, building one nest on top of another, untidy tiers of sticks and leaves and paperbark up to seven levels high.

Ducks nest there, too, eleven natives and the abominable Mallard and perhaps even the rare Freckled Duck. Arthur stresses the importance of this breeding ground and begs for more water to extend it. Fish spawn there, too, in the open lagoons and Victorian scientists are helping with a study of what water native fish need. Every river in the Murray–Darling basin is controlled in some way, most of them by major dams. Water has to be rationed out, and so far farmers come first.

The rationing is complicated. Even the position of a River Red Gum forest matters to some creatures. If western forests languish because of salt or too little water, the Regent Parrot will disappear. This magnificent bird, yellow amber with red splashes on the wing, nests and roosts in Red Gum but feeds on the seeds of Hibbertia in the Mallee. It will not fly more than 20 kilometres from roost to feeding ground.

The great concern in the western Riverina and in all intensive irrigation districts is salt. It is an astonishing experience driving north up the Calder Highway to Mildura. The broad strip down the centre of the divided road is Mallee in a 250 millimetre annual rainfall district. Either side of the road are lush vineyards and orchards. The Chaffey brothers from California began irrigation in 1887 on 100 thousand hectares. The great pump of his own design that George Chaffey had cast by Tangyes Limited, 4.5 metres in diameter, direct-driven by an upright three-cylinder steam engine, pumped 9 million litres an hour from 1889 to 1955. He turned "a Sahara of hissing hot winds and red driving sand" into a garden.

But the huge irrigated areas that developed — the Murray–Darling system supplies water to three-quarters of Australia's irrigated lands — both lifted watertables and negated the natural scheme for handling salt. Natural salt drainage was an essential feature of huge areas of Australia. Our soils are particularly high in salt, especially where ancient seas once covered them. Most of the excess rainwater that bypassed the roots of plants permeated through the subsoil into freshwater aquifers but some of it dissolved salts in its passage, then, because of its different specific gravity, it found its way into special aquifers which conveyed it to the beds of lakes and rivers. Those drains still operate along the great length of the middle and upper Darling–Barwon Rivers — some of them flush water which is twice as salty as seawater. When the river is low, one can see them clearly from the air. They show up as reddish brown stains in the water. As the salt mixes, it precipitates soil particles, so that there is a short stretch of clear water downstream of the springs. Muddy water in Australia usually indicates healthy water, clear water; is often salt.

For the salt drains to operate, it is essential that rivers drop to a low level, not regularly, not even yearly, but periodically for some months. Then a fresh or a flood dilutes the salt which has collected in pools and carries it out to sea. The natural state of our waterways was one of great irregularity. Two hundred odd dams and weirs have changed that pattern to a dependable and unexciting constancy. The breeding of fish, ducks, so many creatures was triggered by the excitement of a wave of new water. More damagingly, artificially high levels of water for years shut off the salt drains by forcing fresh water back up them. Then water tables rise, bringing salt to the surface as has happened over so much of the irrigation areas of the Riverina. The lunatic scheme to bring Clarence River water across the Dividing Range into the Murray–Darling system (one still hears it seriously discussed) would ensure the salting of the entire basin. It would destroy the most productive area of Australia.

The mean level of all rivers is now too high for them to act as sumps. Two hundred thousand hectares of the Murray–Murrumbidgee irrigation has a water table approaching the dangerous 1.5 metre level, the point where surface water has immediate contact with underground water and salt rises to the surface by capillary attraction. No matter what is done, this will increase to 300 thousand hectares over the next five years and ultimately 75 per cent of the land will be affected. Every minute 2.5 tonnes of salt flows over the South Australian border in the Murray waters. Most of this huge quantity now flows into streams with surface water.

If action is taken early enough, salting can be restrained by growing trees to act as pumps and transpire water before it becomes troublesome. In 1985 Lake Toolibin in a farming area in south-west Western Australia threatened to become a salt lake. Grass on the surrounding farms that it drained was already yellowing. The big lake was more swamp than lake. Tea-tree and paperbark grew thickly over most of it and thousands of birds nested in them and among them. The farmers formed a tree-planting group, fenced off the lake foreshores and long wide strips of their farms, then planted trees with some salt tolerance, 100 to the hectare of a mixture of the eucalypts River Red Gum, Flat-topped Yate and Moort, and *Casuarina obesa*, Swamp Oak. In addition they put in sufficient seeds of saltbush to produce 1000 plants to the hectare: Bluebush, Wavy Leaf Saltbush (*Atriplex undulata*) from California and the local River Saltbush (*Atriplex amnicola*). For good germination, saltbush seed needs soaking for a couple of days; *amnicola* which has a more substantial germination inhibitor than most species, responds to a little detergent in the water. Already the grass is green again. The farms and the waterfowl seem to be safe.

It is seventy years too late for early action in the Riverina. Lamberts Swamp west of Mildura shows what has happened and what will happen. Steve Page, with the rural Water Commission of Victoria, took me out there. Before the setting up of the Merbein Irrigation District in 1910 this was a beautiful wetland, a natural drain with trees and shrubs transpiring surplus water into the air — Red Gums, Black Box, Eremophila, Grevillea and clumps of Nelia (*Acacia loderi*) that produces a gum prized as food by Aborigines. By 1930 the trees were dead, their roots unable to lift the salty water. Ground does not like to lie uncovered. Saltbush and other tolerant plants took over for a time, but now the shores are bare, too salty even for Samphire and Sea Blite. All that blooms there are white crystals. The water is now three times saltier than the sea. The lake drains about 65 hectares only, yet the tile drains under vineyards and citrus orchards, together with saline underground water outcropping from elsewhere, pour in twice as much water each year than the sun can evaporate. To stop the lake spreading and salting the whole area, the surplus water has to be pumped out into drains and shandied with fresh water from a delivery channel before it reaches the Murray so that the South Australians do not complain too much when it reaches the testing station at Renmark.

For years the only correction to rising water tables has been mechanical. One engineer recently made a serious suggestion that the Murray should be used as a salt drain and all fresh water carried in newly cut channels. At Wakool, where extensive ricegrowing began in the 1950s, 2000 hectares went completely out of production in the 1970s, yields declined to half on half the rest of the land. Now forty-eight pumps have rescued 47 thousand hectares. They each lift 40 thousand litres of salt and water an hour and force it through 26 kilometres of pipe to evaporation basins covering 2000 hectares. The salt will be harvested: production about 240 thousand tonnes a year.

These basins are not deserts of heavy water and glistening crystals. Australia has ample creatures to enliven its natural salt lakes and many of them have come in. Fairy Shrimps can live in water six times more saline than seawater. They swim on their backs, paddling leaflike appendages and feed on detritus and the algae that have also adapted to salt water. They breed in millions. Their tiny eggs blow about in dust which is probably how they are carried to new areas. The eggs can stand years of dessication, they might even require it. Scientists are reluctant to believe that eggs might last up to fifty years between fillings of Lake Eyre, but it seems that they might.

As well as several species of Fairy Shrimps, there are saltwater beetles and water fleas, the larvae of *Chironomids* or midges, and Seed Shrimps, crustaceans with a bivalve shell. Australia possesses the biggest of these creatures, even though they are not quite 4 millimetres long.

With all this good food, waterbirds flock to the basins and the Department of Water Resources has provided islands for them as refuges and breeding grounds.

But there is doubt about how much of this water is evaporating, how much is seeping into underground water. Perhaps too much of it is recycling. Anyway, there is no room for such schemes to cover all the irrigation districts even if the money were available. Costs are huge. Trees and shrubs now have to do the pumping, millions of them, many more than the land has ever carried. School children learn how to gather seeds, how to germinate them and grow them, how to plant out seedlings, how to sow seed directly into the soil. They no longer put in a tree as a ceremonial event on Arbor Day and forget trees until the next year. They sit on floats fifty at a time with signs they have painted, "TREES FOR LIFE SALT FOR DEATH." They experiment with different methods of planting and sowing and watch the results. Above all, they go home to their parents with theories of how to manage water.

It is astonishing to find that irrigation methods are bad, appalling would not be too strong a word. Farmers do not even make elementary calculations about the length and width of bays, the gradient, and the absorbtion rate of their soils so that they know exactly how much water to put on. They open the gates and what does not run in runs off. Spraying should be a better method than flooding but many of those who spray also work in town, so they turn on their pumps when they go to work in the morning and turn them off when they get home. Often two or three times the correct amount of water is put on. Very few use soil testers to check when to irrigate, not even soil augers, a piece of equipment I was told was vital when I went to an irrigation school at Yanco before we began irrigating at Boggabri in 1956.

Things will change. Farmers in every area are setting up bodies such as the Nangiloc–Colignan Salinity Work Group (one end of the district reverses the Aboriginal name of the other end) to solve their problems with advice from government experts. There will soon be 150 of these Landcare groups covering one million hectares of farmland. Probably every area will need different treatment. Ricegrowing will have to be phased out, despite the fact that our farmers produce the world's heaviest crops. These farmers are responsible for half the problem water. One cannot ration water to the rice plant, it must grow for months in flooded bays.

Generally, there will be a change to drip irrigation for all other crops by those who can afford it — and it may well be that all must soon afford it — and there will be reuse of drainage water on communal lucerne crops and woodlots of species that can take the salt, the boron and the nematodes in it, like the River Red Gum, Yellow Gum and Flooded Gum of the eucalypts and the beautiful Drooping She-Oak, *Casuarina stricta*. These trees produce good timber and good firewood. The woodlots will be valuable in fifty years' time. These are long-term projects. Ordering of water will change, too, for all irrigators. Groups, not individuals, will order bulk requirements that have been calculated exactly.

Delivery channels are a problem. Only 5 per cent of the stock water sent to Mallee farms ever gets there. The rest seeps out on its way and is wasted. The big irrigation channels lose water, too. All of them will soon be flanked by trees, some of them with exotics that are being tested: Date Palms, Carob Bean, *Pinus pinea* that yields pine nuts, *Robinia pseudoacacia* that is used to mop up runoff in Italy's vineyards (it grows so quickly it can be harvested every five to six years, coppicing each time it is cut; it burns green, it yields good fence posts, the leaves are good stock food) and several species of Paulownia, the extraordinary trees from China and Japan. Their wood has no smell, so it can be used to make packing boxes for food, the timber is good for building and it can also be shaved as thin as paper and printed on directly, the roots fix nitrogen in the soil. Agroforestry will become a well-known term. But mostly it will be millions of newly planted Australian trees that will be sopping up excess water in the Murray–Darling basin.

Farther south in Victoria there is dryland salting, as farmers have found in Thomas Mitchell's Australia Felix, the rolling downs of the Western District. As that great explorer moved down the Glenelg River about 50 kilometres to the west of present Hamilton, he noted that "forest land ... now opened into grassy and level plains, variegated with belts and clumps of lofty trees"; the country resembled "a nobleman's park on a gigantic scale", it "gladdened every heart". On his return journey, just south of the Grange Burn that flows past Hamilton, he found hill slopes covered with Silver Banksia, north of the Burn grew Black Wattle, *Acacia mearnsii*, trees 25 metres high with a 10 metre spread. This is another Australian tree named from an overseas specimen. E.A. Mearns, an American botanical collector, described it in Kenya in 1909. It was growing so widely there that he thought it was a native of that country.

On first impression, the Hamilton country is still beautiful. The paddocks are dotted with one of the many inland forms of River Red Gum that do not need flooding, spectacular sprawling trees spaced from 20 to 100 metres apart with trunks up to 2 metres in diameter. Then one realises that these trees are old — 400, 500 years, some older. And they are showing their age: their leaves are small and reduced in number. A few are dead, indeed 2–5 per cent are dying each year and that rate is compounding. There is no regrowth, there are no shrubs, no flowers. This is degraded country. And now there is salting. It is a wet land in the winter and too many of the sponges are gone.

Hamilton began work ten years ago. But Professor Carrick Chambers of Melbourne University thought the story of the good work being done with experimental tree planting was not getting out to enough farmers. The Department of Conservation, Forests and Lands is not good at advertising itself. Chambers worked through friends and approached the Potter Foundation set up by the stockbroker, Sir Ian

Potter, to generate worthwhile projects. Their aim is to set a project going then pull out, not to manage it. The Foundation agreed to put in two-thirds of the money if the chosen farmers put in one-third. Most saw it as an outstanding opportunity and paid half the costs. Between 1985 and 1987 the Potter Foundation established a Farmland Plan and spent about $1 million on fifteen demonstration farms. The results are astonishing: Hamilton is now the tree centre of Australia.

The Milne family of *Helm View*, west of Hamilton, set up their farm under the scheme and they have continued to plant trees. They agreed to hold two or three field days a year, but now they are getting two or three visits a week, many of them busloads of people. Bruce Milne believes their farm originally had about twenty-five big Red Gums to the hectare on it. For Australia, it was unusually well timbered. Sawmills and sawpits worked all about the area, especially at the turn of the century, cutting sleepers and paving blocks for the streets of Melbourne and Adelaide. During the depression years when labour was cheap, farmers had a lot of their remaining big trees ringbarked. There was no fertiliser spreading in those days, no sown pastures. They thought the dead trees would allow extra grass to grow for extra sheep.

When Bruce began planting they were left with fifty Red Gums on 1560 hectares. Four hundred and forty hectares was a recent purchase that had two native trees left on it, both Black Wattles. Now they have 35 thousand trees fenced off on 12 per cent of the farm. Red Gums have grown 4 metres in 4 years. It has made no difference to the stocking rate of ten sheep to the hectare over the whole farm. His aim is 60 thousand trees on 20 per cent of the farm, which he expects will gradually increase the number of stock he can carry. They will eliminate unproductive wet spots and incipient salting, birds and wasps attracted by them will keep down scarab larvae that so damage pastures.

In that country of good rainfall there is no need to fallow the plots to be planted. Bruce rips the ground in late autumn, he sprays at the end of winter with the knockdown Roundup and the supressant Simazine to avoid the seedlings being overwhelmed by Cape Weed or the vigorous introduced grass, Phalaris, and he plants tubed stock in September and October in L-shaped windbreaks as well as along creek banks and in connecting corridors beside fence lines. They get no further attention and 95 per cent prosper. But 40 kilometres of electric fence guard the plantations from stock. There are few conventional gates in any of his fences. Instead, long, hinged, triangular, galvanised piping frames swing up and lift panels of the fence high enough to drive sheep and cattle under, or a tourist bus.

Behind the fences, *Casuarina stricta* demonstrates why it is called Drooping She-Oak. Out in the paddocks these trees are trimmed as high as cattle can reach. There is little evidence of droop. In the plots, the branches sag and leaves sweep the ground.

As we drove round the property, Bruce shook his head at a neighbour's ploughed paddock. Until the watertable lifted, much of this country was farmed, but now most farmers realise that ploughed ground allows too much water in. Until their trees lift it out, none of it should be farmed.

The men who settled this country followed Mitchell's tracks. He had a reputation for happening on good ground wherever he went. At that time the water table was 14 metres down. On 10 per cent of farms it is now just on the surface in low-lying parts and it is one-half to one-third as salty as seawater.

The CSIRO tried to lower water there by growing lucerne and phalaris instead of trees and shrubs. It did not work at all. They grew outstandingly but the height of the water table increased. The dense green growth held back runoff water from heavy rain and it soaked in. But there is an impervious clay layer under the topsoil. How did water soak through that? They dug a deep trench to find out. As well as being well timbered, the farm was apparently one of those places where mixed shrubs grew densely. Rotted roots in great numbers showed up in the sides of the trench and these acted as aqueducts. So the dead roots take down water that the living roots pulled out.

Already Bruce has seen dragon flies flying in mating loops over pools in the creek, Superb Blue Wrens search for insects in the fenced-off bushes. For many years this land has known only the ubiquitous birds like ravens, crows, galahs, peewits, magpies, wagtails and introduced sparrows and starlings, and coarse insects like the scarab beetles in unnatural numbers producing the grubs that eat roots of grasses. His tree corridors connect up with land not so degenerate and allow delicate creatures to come back in.

Neil Lawrance at Balmoral on the Glenelg, 50 kilometres to the north, formed the Dundas–Black Range Corridor Group five years ago.

Via roadsides, laneways, farm gullies, boundary fences, an unused railway line, over recharge areas and down discharge areas they are connecting the small isolated Dundas forest with the bigger Black Range forest 14 kilometres away. Degraded farms act like prison walls to animals.

When I rang Neil he told me he had just come in from "blistering a couple of swages". Swage is not in any of my dictionaries but it is an American term coming into common use here: swells and swages are hills and hollows. It is in the swages that saltwater is likely to outcrop, so they double the width of the tree plantings in them. What Neil had been doing was moving fences out to encompass double widths of planting: it made bumps in his fence line. Width is determined by the width of the boomspray in use, usually about 16 metres. Nobody likes using dangerous sprays, but it is essential to control weeds in that counry or they will overcome the tree seedlings. And wherever a swell is a recharge area, a point of entry for water, they double plant it too and blister the fences around it.

Neil began planting twenty years ago when he found he could easily count the trees on his 1400 hectares: 100 She-oaks, fifteen Yellow Gums, 100 Yellow Box, one Blackwood. He sowed seeds of Blackwood and now he has thousands in an interlocking network of mixed trees. Willow-leafed Acacia (*iteaphylla*), from the Flinders and Gawler Ranges of South Australia, that he planted on slightly salted soil twenty years ago, seeded itself and he now has thousands of them too among Swamp Oak, Kangaroo Paperbark (*Melaleuca halmaturorum*), Swamp Yate from Western Australia and local eucalypts. He puts in many shrubs, too, with the aim of a flowering over most of the year for insects and small birds. We saw newly planted shrubs in a well-treed roadside that was part of the corridor.

Tall Wheat Grass, an import from South Africa, provides substantial stock feed on moderately salted areas. Neil knows of other grasses overseas that give greater promise but they have yet to be tried in Australia. They are going to be needed. In Western Australia, 20 per cent of some shires are already salted. Extension officers are wary of mentioning such figures. They fear it will frighten people too much. A difficulty with feeding livestock on plants that take up salt is that they must have fresh water, difficult to supply in saline soils. Sheep did not do well in the Riverina where both water and feed were salty until freshwater was brought through in channels.

There is wind erosion, too, about Balmoral. Sheep camp on ridges, destroy the pasture and the soil blows. Many hills are bare on top for up to a kilometre. So they, too, are being fenced off and planted and the sheep will camp beside the protection of the trees on more stable soil lower down.

The drive up to the Lawrence homestead is through an avenue of old pines. All through Victoria there are homesick plantations of European trees of 150 years ago, and memorial drives of exotics from the end of World War I. They are of little use to Australian animals but they are now important relics. We saw rows of old Pepper Trees, Weeping Willows on creek banks, avenues of pines and cypresses, even wide rows of the poisonous Oleander leading up to one homestead.

Along the Grange Burn plantings are being made to shelter the Eastern Barred Bandicoot, a delicate creature grizzled yellow-brown with light stripes across its hindquarters. It was thought to be extinct on the mainland until a healthy community was found living in old car bodies at the Hamilton tip. Then the council buried the rubbish. Luckily there were lesser communities along the creek banks. They look for good cover — the car bodies were ideal, but low spreading shrubs will replace them.

Dr Rod Bird of the Department of Agriculture and Rural Affairs is experimenting both with sprays and with direct seeding of trees and shrubs from his base at the Hamilton Pastoral Research Institute. A correct spray rate is difficult, too little is ineffective, too much prevents germination of tree seeds. Some of the new sprays are frightening, dangerous to operators and too chancy since their spheres of influence do not seem to be confined to where they are sprayed. If weeds defy the presowing application, he uses an overspray after germination which also requires nice application of the right chemical or it will kill the new trees. The work with seeds is more congenial. So far he has 100 thousand trees growing in his own experiments on fifty sites and another 100 thousand that he sowed on Potter demonstration farms. He is testing about eighty species for salinity planting — so far Swamp Oak (*Casuarina obesa*) has proved the best — and forty species for farm woodlots. Simple hand sowing of seed mixed with sand gives him good results. Several experiments are designed to find out how thickly to sow. He considers that the clearing of the Silver Banksia that Mitchell passed through has a lot to do with the district's troubles. As well as on the south of the Grange Burn, it stretched in a thick, long, wide belt across hilly recharge areas throughout the district. The first farmers cut down the lot, letting water underground in the cold waterlogged months of July, August and September.

When planting the variable River Red Gum, its provenance is important. Seeds from trees on the Murray or from Lake Albacutya on the edge of the Victorian Big Desert give by far the best results.

In 1982 John Fenton and Roger Young set up "Rural Trees Australia" to design farms. Both are farmers, they knew the problems and some of the answers. First in a great new field, they have doubled their work each year. Jim Sinatra, David Hay and Donald Thomson joined them as landscape architects. The big maps they draw when planning a property are meticulous and consider what will happen over the next fifty years at least. Tree lines will never cut off the wind from a mill, or a new gully wash in heavy rain because water has been turned wrongly. They work as far east as Gippsland, north to Coolah in central New South Wales and west into South Australia. The cost of a plan for a property of average size, about 3200 hectares, is $8000. Big properties cost less per hectare, small properties more. In addition they will prepare land for plantations, supply trees and guards, plant them, even manage them if that is required. They have tree planters for hire and spray units.

They do not overlook the importance of single trees and mark them on the maps, both those in place and those to be planted. A lone big tree can transpire as much water as several in a clump. It produces more leaves and the wind can get at it better to hasten evaporation. But the bulk of tree planting is of corridors and avenues beginning on the recharge areas below hilltops and above the rings of granite that direct water down. A growth of only three or four years pulls a significant amount of water out. When we spoke to the designers they had 50 thousand trees to plant over the next three months in a district where winter planting is possible. As well they were preparing a hundred-year management plan for the Wando River Landcare Group, 120 farmers on 35 thousand hectares. Funding from a new federal body, National Soil Conservation, had just come through. At Robe in South Australia they had established a crop of 14 thousand banksias for the export flower market.

This work is exciting. Farms have never been considered as entities growing over the next hundred years. Most of the planning is about next year's crop or the next wool clip. But what they design will not become sterile completions like a Garden City, they will be frameworks that each district will build naturally from its individual resources.

Murray Gunn, an associate director of the company, has set up his own business, Elsey Seed Supplies, to harvest seed from chosen specimens of local plants, and from other Australian plants growing well in the area. The healthiest seed is always parochial, local conditions impose something on them. Murray also has a breeding pair of Eastern Barred Bandicoots at the end of his lovely plantation of trees.

Like the Riverina, the great Mallee of northern Victoria and eastern South Australia has lost its function of protecting land against salt. There is not enough of it left. Originally it covered 5 million hectares. Farmers and governments saw it as new wheat land and cleared 4 million hectares. A mallee is an extraordinary little eucalypt. It throws up multiple stems — two, three, four, perhaps a dozen or more — from lignotubers, the so-called mallee roots, varying in size from that of a football up to distorted masses weighing a couple of tonnes. Sometimes all of it is hidden below ground, sometimes it swells above it. The rainfall where it grows is 250 millimetres (10 inches) and at least eight species grow thickly. Mallee is a tough plant. It flowers profusely and honey-eaters and lorikeets with brush tips on their tongues lap nectar and pollen and fertilise the flowers with the pollen that catches on the sides of their beaks and builds up into noticeable bulges. Hundreds of species of insects work the flowers, too.

Many shrubs break the sprawl of mallee — grass trees, banksia, hakea among them. A little cypress pine with warty fruit grows in clumps, big belts of Belar run through it and Rosewood, known there as Cattle Bush, grows among the Belar. There are big trees, too, Red Gum and Brown Stringybark mostly. In some springs the wildflower display is equal to Western Australia's, natural gardens on a scale to be seen nowhere else.

This great forest used all but one-tenth of a millimetre of the annual rainfall. The soil is saline. Twenty to thirty species of saltbush and samphire lifted salt out of the topsoil and stored it in their tissues, the

The Darling River advertises its trouble with a thousand-kilometre banner of blue-green algae.
Photograph: Lee Bowling, NSW Department of Water Resources (The blue-green Darling River)

Pilliga Forest: The big, old white Cypress Pine was allowed to die a natural death because of its twisted grain. Photograph: J. van Loendersloot (Out-of-the-way Australia)

North Queensland ventilator, Boer War Rotunda, Charters Towers. Photograph: Elaine van Kempen (Out-of-the-way Australia)

The irrigation channel that should be the Murray River. Note the bare banks and lack of water weeds.
Photograph: Elaine van Kempen (Spring)

The Namoi River at Keewadyin, Boggabri, now little more than an irrigation channel. I planted the poplars in 1958 before I knew the danger of exotic vegetation. Cattle kept them trimmed for years.
Photograph: Elaine van Kempen (The soil that gives me substance)

An especial Kurrajong is about to fall over. Photograph: J. van Loendersloot (The soil that gives me substance)

Dingo. Photograph: Vic McCristal (Song of the dingo)

A magnificent River Red Gum. Three hundred years of floods have built metres of silt about the trunk. Photograph: Elaine van Kempen (Trees, trees, trees)

Trees planted to mop up the underground water that is beginning to lift salt into this depression on Geoff Anderson's mallee farm at Ouyen, Victoria. Photograph: Elaine van Kempen (Trees, trees, trees)

Basking Green Turtles, Bountiful Island. Photograph: Penny van Oosterzee (A voyage of discovery)

Poached Eggs as the sole cover of degenerated land west of the Darling River. Photograph: Elaine van Kempen (The blue-green Darling River)

Poached Eggs, spectacular but useless. Photograph: Elaine van Kempen (The blue-green Darling River)

Hardwood logs to the Casino sawmill, early 1900s. Photograph courtesy Richmond River Historical Society Inc. (The north coast story)

Cane tug pulling three punts to the broadwater sugar mill, Richmond River. Photograph: John McLeod c.1890. Courtesy Ballina Public Library (The north coast story)

S.S. *Aggie* towing a raft of hoop pine logs on the Richmond River c. 1910. Courtesy Richmond River Historical Society Inc. (The north coast story)

S.S. *Irvington* plowing through water hyacinth on the Richmond River c. 1916. Courtesy Richmond River Historical Society Inc. (The north coast story)

A clump of Triodia, the desert spinifex that dies so quickly without fire to rejuvenate it. Photograph: Elaine van Kempen (More a new planet than a new continent)

A cart load of rabbits for a freezing works in Robe, South Australia, 1920s. Photograph courtesy Alec Barrowman (More a new planet than a new continent)

little water that carried down drained salt slowly into lakes like Pink Lakes, coloured by an algae, that produced all of Victoria's salt from 1916 to 1975. Any runoff from sudden storms flowed into freshwater lakes flanked on the east by beautiful lunettes, half-moon sand dunes covered with shrubs.

Farmers came in with mallee rollers, heavy wood or iron contraptions a metre in diameter and 3 metres long that they pulled with bullock teams. Horses would not take the slapping of branches against their bellies. When the broken branches and leaves dried, they fired them, then grubbed out the roots which became famous as firewood. As soon as it was ploughed the soil began to blow away alarmingly. Rain ran straight into the subsoil and raised the water table. Underground aquifers, could not cope with a hundred times more water. Freshwater lakes became salt, productive low ground became salt bogs. Because of clearing of mallee in South Australia, rich land along the Murray will salt in sixty years' time according to scientists who have measured the rate of flow of the underground water.

Geoff Anderson of Ouyen showed us one bog of 250 hectares. It grows a little Bluebush now. His father as a child climbed tall paperbarks there, known locally as Moonah. The very first wheat crop caused the salting. But Geoff has a big farm, the area is not significant. He grows a lot of wheat and barley with big machinery but he has changed his farming practices, spraying strategically with low concentrations of chemicals then grazing the dying plants. He aims at one cultivation only before sowing. It gives less time for wind erosion, less water gets too far underground. And he is planting trees, thousands of them. From the 1960s to the mid-1980s many Mallee farmers planted Western Australian trees. Some blew down too readily, a lot died after twenty years. A Red Gum subspecies from Broken Hill, Silverton Gum, is proving outstanding, reaching 5 metres in four years. He sowed Tall Wheat Grass ten years ago and is experimenting with saltbushes to get sheep feed where salt is outcropping.

Brian Keam, Assistant Shire Engineer at Ouyen, drove us out to Murrayville near the South Australian border, where thirteen years ago the water table had reached street gutter level. The hotel had to pump water out of the cellar, the garage out of its greasing pit. The shire planted eighty Sugar Gums, boring into the limestone that lies just under the surface to do so; in addition, they persuaded other authorities to reroute power lines and telephone lines so that the street trees did not have to be lopped. One Sugar Gum can pump out 1100 litres of water a day. Now the water table is 4 metres down.

Ian Thomson, teacher of Agricultural Science at the secondary college, is making every child aware of the fragility of the land. They test soil, identify plants, collect seeds. They are growing trees, planting them, experimenting on salted areas by running lines of plants from high on the bank out into the salt to see what concentration they can take. One plant they are trying is Saltwater Couch, which can stand irrigation with pure seawater provided it gets a flushing with freshwater occasionally. But much salted land is many times saltier than seawater.

The wondrous Mallee Fowl was almost cleared with the mallee. The male maintains the temperature of the sand and leaf mound he builds for the female to lay her eggs in by testing it with his tongue and opening it to the sun to warm it or closing it to cool it. Those fowl left in the Big Desert and in the remnants of the Mallee have been separated, so the whole community — the Shire, Landcare Groups, individual farmers, schools, Conservation Forests and Lands, Department of Agriculture — have combined to plant a corridor 60 to 100 metres wide, 30 kilometres long to connect the surviving birds.

Several machines can be used in tree planting. A shovel and a bucket are too slow to handle thousands. A tobacco planter with seats for two operators can put in 1000 seedlings an hour, other machines with one operator about 2500 a day. All the machines have a V-shaped share to open a furrow, a chute to drop the trees down, a reversed V to close the furrow and angled rubber wheels to firm the soil each side of the seedling. A necessary guard stops the operator being pelted with clods off the tyres of the tractors drawing the machines.

To get saltbush growing in already salted areas, mouldboard ploughs are used to make ridges about 70 centimetres high so that the plants can be kept out of the saltwater until they are established and able to cope with salt. Mouldboard ploughs which turn the sod completely are also being used instead of sprays in some areas. Drawn by powerful tractors they bury surface weed seeds too deeply for them to germinate.

Methods for direct seeding range from broadcasting by hand, to hand-pushing vegetable sowers, to sowing by special machines that in one operation raise a low mound, roll a furrow into the top, drop in seed and cover it with vermiculite, a damp spongy mineral. Where wind causes erosion, a light spray of bitumen anchors seed and soil; in cold climates it also attracts necessary warmth. Rod Bird in several of his experiments uses a scraper 20 centimetres wide to move spray-contaminated soil, then he drops the seed mixed with sand, bran or sawdust on the bare ground and dusts it over with a drawn bag. Some seeds have to be soaked in boiling water, all need varying periods of damp weather to germinate. He aims at one tree per metre of row and now has a good idea of how many seeds he has to sow to achieve that. Paperbark needing to be kept moist for two or three weeks gives only 0.2 per cent success, Blackwood over 20 per cent, so with that species he has to sow only four times the number of seeds he wants to germinate.

The Shire Engineer at Albany in Western Australia, where rain is usually predictable, takes off weedy topsoil for landscaping elsewhere with a big scraper scoop, then scatters seed on the hard subsoil with great success. In Western Australia, too, a group called Ribbons of Green is running a corridor of trees 600 kilometres from Perth to Kalgoorlie. Since last winter, 1000 volunteers have planted 150 thousand trees in five shires and now other groups are forming. In South Australia, Koala Green planted 15 thousand Candlebarks and Manna Gums to give koalas more room in the Mount Lofty Ranges. As well as supplying the best leaves for koalas, these trees are exceptional. Manna Gums grow tall and the bark hangs from them in long, twisting ribbons. When Candlebark strips in the spring, the new bark glows brilliant pink and red.

Tasmania plans to regenerate native bush in those areas that seem now to be wholly English. Exotic gorse is the only shrub and it is a proclaimed pest. One has to look into the distance to see a eucalypt. New South Wales is planting the slopes of the Dividing Range where dieback has killed so many eucalypts. Old trees well-spaced in paddocks were attacked repeatedly by immoderate numbers of insects — Christmas Beetles, Leaf Beetles, Sawfly larvae — until they no longer had the energy to grow new leaves. There are several reasons for the numbers of insects. The roots of imported weeds and grasses are particularly attractive to the grubs of Christmas Beetles, there are no shrubs in the pastures to attract birds and predatory wasps, the great number of eucalypts now on Crown land, many planted more than 200 years ago, sustain many more insects.

The first attacks seem to be the most devastating. Armidale which thought it would lose all its trees in the 1970s is now suffering no more losses, indeed some trees have returned from the dead. Boorowa on the southern slopes is now fearing that it will lose all its trees. For a time New England landholders planted exotics as the only safe trees, now they are replacing the natives they lost and building on their numbers. The Armidale Tree Group has planted 30 thousand trees since 1983, another 30 thousand native trees are going on to 10 Walcha properties this year alongside experimental plots of direct seeding. Similar plantings are taking place all the way down the slopes.

On the central and northern coasts of New South Wales, groups have formed to save rainforest relics, some as small as 8 hectares of the Wingham Brush, which have been overrun with privet and Camphor Laurel and extraordinarily vigorous vines, mostly from South America, like Cat's Claw, Potato Vine, Balloon Vine, Madeira Vine, Wandering Jew. They climbed trees 30 metres high and smothered them. A photograph of the Wingham Brush before they were torn down shows it draped in green shrouds. The workers moved in with bush knives, axes, tomahawks, chainsaws, hooks on long, extendable aluminium poles and the poison Roundup. It needs regular maintenance, but it is now a healthy forest again.

Work under Greening Australia on the North Coast is planned to correct riverbank erosion, landslides, and the decline in production caused by rising soil acidity, as great a problem in many areas as salt. Wide plantings of both dryland and rainforest trees around paddocks will lift neutralising elements from underground and distribute them through leaf fall and insect droppings. Where they can restrain developers they will replant mangroves, the breeding grounds for so many fish and crabs.

In north Queensland, a new group in the Mission Beach–Cardwell area, south of Innisfail, has made a submission to Greening Australia for help in planting rainforest for cassowaries, now in hundreds instead of the previous thousands. The life of a rainforest depends on them: the seeds of many plants are processed for germination in the gut of cassowaries and they grow in the droppings, even the beautiful Fishtail Palm with seeds so caustic they blister human fingers.

There has not been the same degradation in western Queensland as in New South Wales, probably because rabbits did not reach there in such numbers. The grasslands still have much of the original cover. There will be planting near Warwick on the Darling Downs where salt is beginning to show and on big areas out of Rockhampton.

Greening Australia, which is organising the tree planting, set up as a non-profit body in 1982 during the International Year of the Tree. Until 1999 it has a guaranteed budget of $40 million which will be supplemented by state governments, local councils and individuals, both corporate and private. Hundreds of groups with names like Men of the Trees, Trees on Farms Inc., Tree Planters and The Deniliquin Tree Action Committee will carry out the work.

In addition to planting for reasons already given, they will work with the Water Conservation and Irrigation Commission of New South Wales to plant wide bands along denuded river and creek banks to slow floods and act as silt traps. The Azure Kingfisher, for example, has to be able to see its prey in water up to a metre deep if it is to survive. Even European Carp that stir the bottom when feeding upset that bird. Greening Australia hopes to persuade the Forestry Commission of New South Wales to reduce the size of exotic softwood plantations to 500 hectare blocks that are broken by connecting corridors of native bush, and to plant an understorey among the pines of whatever native shrubs will tolerate them. It will hold workshops in all states to advocate the growing of woodlots from town sewage sprayed on as mixed water and fertiliser, as is already being done at Loxton in South Australia and Shepparton in Victoria. Trees use twice the water that the sun can evaporate.

Not all the work will be done in the country. Gus Sharpe of Greening Australia aims to persuade owners of industrial sites in Sydney to grow trees. So many have waste dead corners that could be brought to life.

There are aspects of tree planting that few have yet considered. When seeds are being collected from outstanding trees, it ought to be worthwhile to take soil from under the tree — only a little would be needed — in the hope of transferring the spores of symbiotic fungi when it is added to the potting mix. This fungi encloses young roots, protecting them, even feeding them, and makes a remarkable difference to growth. None is likely to be present on degraded land. Collecting such soil would not work with all species: River Red Gums, for example, drop an inhibitor with their reject leaves and bark that prevents their seeds germinating under them where they would have little chance of surviving.

Once trees are growing in plantations, once the wild growth of introduced plants that follow fencing-off is subdued, still more can be done. Even in those districts that still have plenty of trees, the shrubs are missing. There were not enough shrubs in all the plantations we saw. Trees pump water, but shrubs and flowers attract birds and insects.

In a healthy woodland, birds eat about half the insects produced each year, they clean up about 30 kilograms of them per hectare. The little birds, the pardalotes, thornbills, wrens, flycatchers, eat the little insects, cuckoo-shrikes, kingfishers, and the big honeyeaters, like the Noisy Friar Birds that switch to insects when nectar gives out, eat a lot of Christmas Beetles. Several years ago, heliothis caterpillars had reached damaging numbers in a 35 hectare crop of linseed on our farm. I ordered the spray and the plane and feared for the fish in our dams. Then I saw a flock of Black-faced Cuckoo Shrikes feeding in the paddock, no more than thirty. I cancelled the spraying and made daily check counts of caterpillars. In less than a week the birds had brought the infestation to a safe number and they were continuing feeding. We stripped a good crop. Those birds had boundary trees and paddock trees to work from. They would not have come to an open paddock.

The size of a plantation is important. Five-hectare plots attract three times more birds than 1 hectare plots. Although the birds are needed to eat insects, normal numbers of insects are as valuable as the birds. Not only do their droppings, known as frass, recycle an astonishing amount of nutrients, but some insects prey on damaging insects, like the Hairy Flower Wasps that lay their eggs in the larvae of Christmas Beetles and other scarabs. The adult wasps are nectar feeders, favouring Sweet Bursaria and Silver Banksia which grow readily. The female is wingless, so the winged male carries her out from her favourite shrub to find a grub, then he carries her back again after she has laid. Like the Regent Parrot, there is a limit to the distance he can fly — in his case, 200 metres. So ideal paddocks to attract birds and insects, a size that allows for most considerations, are of 40 hectares, 1 kilometre long by 400 metres wide with a 20 metre wide timbered corridor around it. Much bigger paddocks are needed where Black Crickets are a pest. About Kerang in Victoria they come out of protective cracks in the ground ahead of irrigation water and mass in millions on the channel banks. Straw-necked Ibis feed on them, a big flock can keep them under control, but they like open spaces and seldom visit little paddocks.

When everything else seems to be thriving, the flowers can be restored: peas, the terrestrial orchids, the Garland Lilies, the Creeping Boobialla, so many. It would be valuable to bring back Murrnong, the delicious milky, yam-like root with a coconut flavour to plantation edges (it will not grow in forests). Aborigines ate a lot, the women dug up big areas collecting it, and it was the main diet of the Long-billed Corella. Sheep ate the tops then dug for the roots, rabbits cleaned up what was left. The Long-billed Corella almost disappeared, then Onion Grass, an introduced member of the Iris family, spread over the degraded land and the remaining Corellas found its corms good to eat. They bred locally to numbers not previously known.

And there are spiders that, like the flowers, would have too far to travel. Those whose young disperse on gossamer parachutes will find their way in; many species will have to be brought in from where they still exist.

Different parts of Australia have very different problems. Scientists from the CSIRO are working in the west of New South Wales and Queensland, in central and northern South Australia, in Western Australia and the Northern Territory under the National Rangelands Program. I prefer to call that huge area the drylands. Rangelands sounds like a translocation from the American prairie. Almost a third of it is severely eroded — that means the greater part of Australia. Wind and water began to move the soil when rabbits bared the country in the 1890s; in the 1940s it was thought they would eradicate the last stands of Mulga, the acacia that grows in groves wherever runoff water collects. In the east of the region it is a small tree with silvery grey phyllodes about 20 millimetres wide. It diminishes as one travels west until it is an olive-green shrub with phyllodes like hakea needles 1 millimetre wide.

When myxomatosis lessened the numbers of rabbits, Mulga came away again but in many places it came too thickly, occluding the grass, and erosion continued. All this country has been overstocked with sheep. One cannot condemn the graziers. During the push for Closer Settlement after World War II, big holdings were cut up into areas calculated to run 7000 sheep. It was a clause of the agreement that the land be fully stocked within three years. The carrying capacity, by a piece of government lunacy, was reckoned on so-called normal years with droughts excluded. In that country especially, droughts are not a calamity. The soil depends on them to recoup itself.

The latest recommendation is that no more than a quarter of the forage grown each year should be eaten. That allows the most valuable native grasses, those like Mulga Grass that are almost gone, to regenerate. It also allows stock to be carried through a drought without devastating the country. So properties must become even bigger than they were before Closer Settlement, and if the rabbits now building up again can be once more brought under control, the country will recover.

East of Cobar, driving towards Nyngan only a few days ahead of the flood rains, Elaine and I came into country devastated by trees, a forest with hard bare ground as understorey. "The more trees the less grass, the more erosion, both wind and water," Warwick Date, agronomist at Cobar, told me. It is part of a huge area stretching up to the Queensland border where trees and shrubs have sprung up in inedible masses: Bimble Box, Red Box, White Cypress Pine, Mulga, Budda, Hopbush, Turpentine, Punty Bush. These scrubs always grew thickly in narrow belts and small clumps. As such they were valuable and some of them are beautiful. But spread over hundreds of thousands of hectares that explorers described as open woodland, they are useless even to most native animals.

I gave the reasons for such growth in *A Million Wild Acres*. Most of it took place in the 1880s; this present eruption began about 20 years ago and is increasing. How it can be dealt with depends on the rainfall and the monetary value of the country. A landholder cannot spend $10 a hectare on land worth $4. The simplest way is to spell a paddock of 2000 to 3000 hectares for a year to allow grass and leaf litter to build up then to burn it when it is hot enough to give a complete leaf scorch. It is not necessary that flames carry right to the topmost leaves, but these shrubs are so hardy that half a dozen leaves preserved on the tip of a high branch can resurrect the shrub. The spelling of the paddock has to continue for one year, two years, until there is enough good rain to germinate grasses and herbs. The seeds of most Australian plants have astonishing endurance.

On land that will bear the cost, land that will run a sheep to 1 hectare when it is rescued instead of to 4–5, 10 tonne chains 120 metres long can be towed between two caterpillar tractors to tear trees and shrubs out of the ground. Disturbed soil, protected by a mulch of dead scrub, is a good medium for the wanted plants to grow in. Some landholders are experimenting with an even more expensive method, blade ploughing, which involves towing an implement which is nothing more than a massive knife designed to run about 20 centimetres under the ground. This takes a lot of power. The tractor that can comfortably pull its half of a 120 metre chain works hard to pull a blade plough 4 metres wide.

It is a bonus for the expensive ripping of rabbit warrens that saltbush and grasses immediately colonise the plots of loose soil. There are millions of warrens in western New South Wales that will have to be ripped at a cost of about $6 each. Landholders have to work together on such projects or else properties reinfest one another. Firstly, poison is laid to lessen the number of rabbits, then packs of dogs are used to frighten the remaining rabbits underground while tines hauled by big-tracked tractors crumble the burrows about them and smother them. If not dealt with, rabbits will negate the tree program.

A parvovirus now killing millions of wild and domestic rabbits in China and Europe might be of use in Australia, but because of the little money granted to scientific research, it will take at least five years to make certain that the disease is species specific and safe to introduce.

Kangaroos, as well as rabbits, are destroying the big Hattah–Kulkyne National Park in north-west Victoria. It has become a wasteland dominated by exotic annuals. Fenced-off experimental areas make a quick partial recovery, native shrubs and grasses grow. There is no money to eradicate the rabbits and Barry Rowe, then Minister for Agriculture and Rural Affairs, had a nervousness about the kangaroos. He would not legislate to make their excellent flesh available for human consumption, he would not even sanction their destruction. The park was established as a sanctuary for rare plants which will disappear. And although no money is available for the study and maintenance of any parks, 775 thousand hectares, an enormous area, has been recently added to the Mallee and desert parks.

These acquisitions are sadly degraded, but shutting them up will not restore them. The answer was to redesign the properties with wide belts of connecting shrubs so that they could be grazed sparingly without damage. That way there are people whose interest it is to keep plant and animal pests under control. And it costs government nothing.

In some of these parks there are big areas of a vigorous *Triodia*, Porcupine Grass, that is not being burnt. Without fire it becomes a desert of prickly dead grass that feeds neither insects nor mammals, the disaster that overcame several desert animals when Aborigines stopped burning.

Kosciusko National Park is a disaster of exotic blackberries and St John's Wort and dense native shrubs, not the open grassland and Sphagnum bogs it has been for centuries. The spongy Sphagnum moss and associated plants guarded the purity of mountain water. The bogs released it in clear trickles. Cattle tramped them into stinking mud, their tracks caused hillside erosion. The answer was to fence off the bogs and the easily damaged country with solar-powered electric fences and to continue the seasonal grazing where it was safe. I have heard the argument that the present dense growth will begin to open up again after about fifty years, that grasses and trees will replace the shrubs. That could happen if the growth were of native plants. Blackberries and St John's Wort do not know the rules, and not even native plants have had much experience in looking after themselves. Aborigines managed them with fire for about 100 thousand years.

The coming of white men with cloven-footed mammals, shovels, hoes and wheels was the second disaster to the nature of Australia. The first disaster was when Aborigines learnt to exploit the fire readily available from volcanoes. The land was too wet then for lightning to have much effect. What they did to the land must have initially looked like devastation, what finally happened was a marvel.

Our own disorder shows up so clearly because it is so new, but we now have the will and some of the knowledge to correct it. Over the radio during our trip we heard a comment by the Tasmanian spokesman for the Australian Conservation Foundation. "It is loggers that have to justify their actions, not us." Anti-loggers are accountable, as are loggers, but neither side has begun the vital scientific study of our forests. Shutting up some forests will kill them, logging will keep them. Nobody knows which they are and it is regrettable that *Our Country Our Future* makes no provision for more money for scientists. The future depends not on politicians, not on public goodwill, but on the knowledge that science brings.

A voyage of discovery

I saw a Chestnut Rail, a rare, shy, crab-eating northern Australian bird, and several Rose-crowned Pigeons, usually lost in the rainforests. I saw an island taken over by Brown Boobies for nesting and we counted 3616 big white chickens while 10 thousand of their parents and their last year's progeny glided low and quietly over our heads. We saw turtles swimming, turtles mating, turtles laying, turtles basking by the hundreds in shallow rock pools. I grew accustomed to going over a ship's side down a rope ladder into a bouncing dinghy at any hour of the day or night. On sandstone shelves beside the sea I watched little rock-wallabies, perhaps a variation of a known species or even a new species. We found lizards and rats in great numbers that escaped what identifying keys we had. We found rare plants and unexpected plants; and adjacent islands that were distinct and different worlds. What I saw with scientists and adventurers during the month of July 1992 were the early stages, the 7000-year beginning, of the development of a group of islands that will behave like the Galapagos Islands and become a living museum of evolution if no one interferes with them.

Some of the islands are so distant and so little known — some of them do not even have names — that it seemed original exploration. To keep a sense of proportion we had to keep telling ourselves that all these islands once had Aboriginal or Torres Strait Islander names, some had Dutch names before European settlement, many of them had been named several hundred years ago by Macassan trepang fishermen from Sulawesi.

But through scientific eyes it was almost original discovery. Some of the islands had never been examined for plant and animal life, on others Robert Brown, the great botanist, had made the only collection of plants when he was up there with Matthew Flinders at the end of 1802.

Penny van Oosterzee and Noel Preece of Desert Discovery, Alice Springs, organised the trip. They are world leaders in such fascinating opportunities which are becoming known by the ugly name of ecotourism. They make it possible for scientists to get to out-of-the-way places for vital research by selling berths to those who wish to accompany them. The money available for science in Australia is calamitously low. Dr Colin Limpus, the authority on turtles who accompanied us, made discoveries on the trip of world-wide significance, yet he had been trying to raise the money to get into the area for twelve years.

We used the right ship for such a voyage — *One and All*, commissioned in South Australia for sail training in April 1986. She is a brigantine, which means that she is two-masted, square-rigged on the foremast, fore-and-aft on the main. Our finest timber went into her, Narrow-leaved Ironbark for the enormous keel, green Karri for the ribs, Tasmanian Blue Gum, Huon Pine. On the deck a change in level of the handrail of Celery Top Pine is joined port and starboard by a glorious sweep of a cast bronze dolphin across the gap. The crew polish them proudly. The slack ends of 84 halyards are each coiled in the same way and hung clear of the deck on their belaying pins so that they can be handled easily, even in the dark. The proudest tall ship of last century was not fitted out better. We did have advantage over Cook and Flinders in a subsidiary diesel engine.

We boarded ship at Cairns, we sailed all night and anchored early the next morning at Cape Tribulation. "Stand by for'ard!" The mate's voice has to carry from the helm aft to the winch in the bow.

"For'ard standing by!"

"OK for'ard, starboard two."

The seaman so ordered lets out the starboard anchor and two shackles of its heavy chain. Each shackle of 15 fathoms (30 metres) is marked by three painted links. "One!" calls the seaman as the first red links appear, and another seaman gives one toll of the polished bell.

We stopped at Cape Tribulation to see Hugh Spencer and Brigitta Flich who established a research centre there four years ago to study fruit-eating bats, which also means an intensive study of the rainforest that supplies their food. It is the only study being made of lowland rainforest, yet they have minimal resources to work with.

It is an important area. Second-rate farms abandoned forty years ago are rapidly becoming first-rate rainforest. Several species of *Melaleuca* grow there, some of them paperbarks on a magnificent scale, 35 metres high and 2 metres at the butt. Some twenty-three species of Mangroves thrive in the tidal zone and display the extraordinary shifts they have adopted to get the oxygen which is excluded from the mud they grow in. Aerial roots stick up like stakes or like bent legs and bare knees, the roots of other species run out horizontally, one over the other, from a few centimetres to a metre above the mud. They twist, fuse, send down anchors till they make a truly formidable network.

Huge slides from Mount Sorrow to the west have remodelled the area for thousands of years. It sometimes rains at the rate of 250 millimetres an hour for two or three days at a time and great slabs of the mountain break loose, bringing down soil, rocks and plants — instant clearing. There is no subsoil there; the surface is supported by big unstabilised rocks. Creeks run on the surface for a few weeks only, then they thread through the rocks underground.

The lively new rainforest supports many bats, but they are not easy to study. They leave areas for no apparent reason, return in numbers at what seem inopportune times. Radio transmitters fitted to them do not last long enough to record their distant and erratic movements, they resist the bigger solar-powered transmitters and tear them off. Hugh Spencer is now putting on devices that emit tiny flashing lights which infra-red nightscopes can pick up. He suspects that bats are the chief pollinators for many rainforest species, especially the spectacular *Barringtonia racemosa* with long drooping racemes of big white flowers. Bats are more important than anybody has ever realised.

There was little flowering or fruiting in the rainforest when we were there, so Hugh put up his trap nets in a neighbour's commercial orchard which is producing breadfruit, mangosteens, star fruit, rambutans and pomelos, welcome offerings to many bats.

Too much disturbance by our visit discouraged feeding that night, but Hugh netted six Tube-nosed Bats, some of which he had previously tagged. They are delicate creatures with a 16 centimetre wing span and strange tubular extensions to the nostrils for which there seem to be no reason. Perhaps they once had a different diet. They now use their sharp teeth to eat fruit like flying-foxes and a short nose would not snuff up any of the fruit. They try to bite when Hugh first picks them up to measure and tag them so he settles them down by feeding them honey and water with an eye dropper. Their wings are a bright cinnamon with black and yellow markings along the fingers which spread them. They sleep with wings folded about them so that they look like dead leaves.

It was a long day, that first extraordinary day. I began it at 3.30 in the morning when I was called to go on bow watch. It was intended to finish about 6.00 p.m. but the seas had risen a bit during the day and the rubber dinghy from *One and All* could not get into the beach where we were. So the ship had to move to quieter water and we had to walk 12 kilometres to meet it. I had spent several months getting fit for such expected eventualities so I found it no trouble. We got back for dinner about 10.00 p.m.

Without sails to steady her *One and All* rolls at anchor. The timbers work against one another, they scratch and squeak and growl like live branches grinding together. The bilge water slaps and the deeper water of the freshwater tanks slops and gurgles.

We ran up to Cooktown in a slight sea with a 28 knot wind behind us and took down the upper topsail before the change of course into the harbour. "In its lifts" cried a deckhand as the sail furled on its yard and brought its weight off the halyard on to the ropes strung from the top of the mast to the end of the yard. The wind tore at whatever part of the sail it could reach, making it flap noisily. Two seamen raced up the shroud, their bare feet hardened to the steel ratlines. They side-stepped out along the yard, supported in the stirrups by the footrope slung under it. One went to the far end and stood on the Flemish Horse, the outermost under-loop of rope. Throwing their arms over the sail, they folded and rolled it, then untied the gaskets (lengths of rope 1.5 metres long), left tidily in place at intervals on the yard, and roped the sail down.

These men were about 25 metres — nine storeys — up, on a rocking yard. Many of the high lines wear baggywrinkles, also called bag-o-wrinkles, to stop the chafing of sails. Old manila rope is spliced on in the form of a chimney sweep's woolly pup.

All the sails had to come down before we anchored out of Cooktown. The tide was not right for us to get into the harbour so the captain sent the dinghy on a rough trip to pick up Jeff Miller, one of the members of turtle research in Queensland National Parks and Wildlife.

Jeff came to Australia as a teacher in 1975, one of the last Americans invited to solve a teacher shortage. He had gained a degree in zoology at the University of Montana where he studied during the winter when the snow was too deep for the packing that paid his way there. For ten years he worked the summer months for an outfitter who organised camping holidays in the mountains. Jeff took campers in by mule and packhorse.

In 1954 he bought a penny turtle as a pet. His mother said, "If you are going to keep that creature you have to learn everything there is to know about it." He is still learning.

The next morning we left anchorage off Cooktown in an unplanned hurry to answer a Mayday call from a yacht. She had lost her mast in a squall, a rope had fouled her propellor, the anchor would not hold and she was drifting fast towards Conical Rock, about 20 nautical miles north of Cooktown. The woman who went the message catalogued the disasters with matter-of-fact calmness. But a prawn trawler which was nearer and faster made a successful rescue.

Before we began turtle hunting we had a good look at Stanley Island in the Flinders Group off Princess Charlotte Bay. It has a sad Aboriginal history. After an occupation of at least 2500 years, vicious Christians in the 1930s moved all the Aborigines off the island to die in mainland reserves.

They left behind remarkable paintings on a well-protected wall. Most of them are of red ochre outlined in white: turtles, frogs, dugong, crocodiles, birds, stingrays and other unknown shapes. Most interesting of all, next to a three-masted sailing ship, are paintings of two craft that predate European shipping. A narrow crevice under the paintings once held brightly painted turtle skulls which commemorated the first kill of young initiated warriors. They were all in place in 1926 but souvenir-hunters have now taken them all.

We found the droppings of wild pigs on Stanley — a nasty find — and, surprisingly, the droppings of some species of kangaroo. None are known to be there. We did not see any of them but I collected the few droppings that I could find to be processed in the hope that a grooming hair or two will reveal the species.

The big Flinders Island next to Stanley has an Aboriginal burial cave on it. We had a long scramble around mountainsides to find it. Grisly souvenir-hunters have taken a great deal from there, too, and National Parks and Wildlife have now fenced it securely. Only two whole skeletons and a few scattered bones are in place. The painted bark that once sandwiched them has rotted. It is not really a cave, more a high, deep recess in a sandstone wall. It looks out over sea. Some mainland Aborigines still take their dead out to islands and deposit them on such lovely shelves of rock where they have an everlasting view.

Flinders and Stanley Islands have a dry climate. Like all the rocky islands close to shore, they are the tops of mountains partly drowned when the seas rose about 7000 years ago as the last Ice Age ended. The wondrous coral reef is built on what were the hills fringing the coast. The twisting shipping lanes, the narrow passes leading out to the Coral Sea, mark the course of ancient rivers and creeks. The many reef islands are sand and dead coral swept together by waves. Initially they grow what seeds the sea brings them. The seeds of these vigorous colonising plants can survive three months in the water. Birds bring in more seeds. Those that can tolerate the conditions grow. Eventually they sort themselves into spheres of capability and each island becomes a separate ordered world.

The reef and the islands on it gave great modern opportunity to turtles and seabirds. Seven thousand years ago there were few islands to nest on, few beds of seagrass and algae to feed on.

We saw our first turtles on the big Corbett Reef about 15 nautical miles north of Stanley Island. Jeff Miller wanted to get some idea of the numbers feeding there. It is not a true island. High tide covers it, low tide exposes patches. We approached it cautiously. The southern sides of all these reefs and islands are classed as "steep-to", a strange old term meaning a sudden rise. In a few metres a reef with half a fathom of water on it jumps out of a 25-fathom bottom. Islands and reefs shelve to the northward, sometimes for considerable distances.

A thirteen-minute run with the dinghy over a specified area produced a count of seventy-two Green Turtles. Jeff was delighted. It was many more than he expected. But the next run counted only three. It was not an error of the counters. Startled by the sound of the outboard, the turtles came to the surface and heaved themselves along the water at astonishing speed. Shells a metre or more long and almost as wide could not be missed. Apparently the turtles feed together like grazing sheep. Still, it was altogether a satisfying count. The tide was falling and we dared not venture farther on the reef. The risk of getting stranded was too great. We set sail for islands to the north.

Night and day bow watch in these busy seas is fascinating. The watch is kept as an intelligent assessment of what the radar has usually seen long before. Wrecks on almost every reef show what has been happening to the careless and the unlucky for hundreds of years. By day there are land, ships, unlit beacons to be reported, by night there are lights.

A light first appears on the horizon as a lume, the term used for a faint glow, and you go aft to report it to the officer on duty, designating its position port or starboard by eight segments of 12 each: perhaps "Lume starboard 2". The lume might disappear without revealing itself if it is a ship on a different course, or it might become a flashing beacon, a line of six prawn trawlers working together or a big ship that becomes many lights then a definite shape which disappears south in reverse order. There was a "strong breeze" behind us and a "slight sea" on that run, which means a wind of 22 to 27 knots and waves 0.5 to 1.5 metres high. Spray broke over the bow, water sometimes ran through midships.

Back on helm duty, steering due north but holding the wheel about 10 degrees to port to counteract the turning effect of the sails, I could sometimes catch the astonishingly normal conversations big ships have with one another when calculating their passing procedures in narrow channels.

"I've been watching you coming down, so I've moved to the west of the channel."

"Yes, thank you. That's what I thought you were doing. I'm running down on the east."

We anchored in quiet water off Douglas Islet well up north of the Great Barrier Reef Park. Another Mayday call set us moving suddenly the next morning. Our captain, Mitch Maciupa, had picked up an excited and incoherent voice telling whoever was listening, "Help! This is a fishing boat. I'm sinking, I'm sinking. I'm at Crocodile Rocks [a trawlers' name], we're sinking. The engine room's half full of water. It's halfway up the side of the engine. I don't know where the seacocks are. I need a pump to pump it out. Will someone come and help me? I'm going to turn over soon."

Iron Carpentaria, a big ore ship working to Weipa, picked up the signal, calmed the captain down, found out that he did not know the call sign for the boat but that he did know exactly where he was — he had a global positioning system — and relayed the message to the Marine Rescue and Co-ordination Centre at Canberra.

Mitch checked his charts, worked out that we could get to him in little more than an hour, went on the air and offered to help. The trawler captain did not expect to be still afloat by then, would we please watch out for an aluminium dinghy with three people aboard?

The trawler looked normal as we neared it. I was watching it, wondering if it were indeed the one in trouble, when with startling abruptness, it rolled over and floated keel up. She was the *Fisco 1*, four nights out of Cairns with a relief captain and a new deckhand. Rosemary Tomlin, the cook, had joined the crew in a Cairns pub. She told me what happened. "I was out the back sorting prawns. I didn't do much cooking. I was sorting prawns. It looked like being a good night, twenty boxes coming up. I heard this 'chug, chug, chug', I looked up. There was a bloody big boom coming straight at me. I ran for the wheelhouse."

Another trawler had overrun them and swiped them in passing. Both captains were out the back sorting prawns. The offending trawler kept on going. A broken rope fouled the propellor on *Fisco 1*. Michael Gottschling, the captain, cut his engine, threw out the anchor, and the three of them spent hours repairing nets and boom. Then they went to sleep without checking below. The collision had torn away the outlet hose from the generator cooler. The big blast freezers on these trawlers work at - 40°C. and it takes big water-cooled generators to power them. The pump for the cooling water had continued to work, pouring the outflow into the engine room instead of back into the sea.

We lowered the rope ladder to the three shocked survivors and took their dinghy in tow. A helicopter sent down from Thursday Island by the insurance company picked them up from Douglas Islet.

The high tides were too low for turtles to come over the reefs to lay on Douglas or Milman Islet to the north. Douglas supports one of the few mature forests of *Pisonia grandis*, a tree named by Robert Brown from rather stunted specimens on an island in the Gulf of Carpentaria that has become known as Pisonia. On Douglas they are 24 metres tall and 3.5 metres in diameter at the butt. A thick layer of rotting leaves covers the mixture of sand and dead coral that they grow in. The trees look rather bedraggled. The big heart-shaped leaves are torn by wind, painted white with the droppings of thousands of the Common Noddies that roost in them. When the birds are taking off to feed in the early morning or returning at dusk, there is a constant heavy shower of excrement which accounts for the degrading name "Piss-on-yers" for this extraordinary tree. It has a savage way of spreading its seeds and causes the death of hundreds of seabirds, especially of White-capped Noddies which nest in them, making neat platforms of wilted leaves cemented together with excrement. The nesting of the birds and the ripening of the trees' fruit occur at the same time, and the club-shaped fruit exude a particularly strong resinous glue from warty excrescences. This sticks to the birds' feathers and, although it does not do much harm if only one fruit breaks away, often a whole spike tears loose which can fasten a bird's wing to its body or glue its tail feathers together. The bird flutters helplessly on the ground until it dies of starvation.

It is not a big forest on Douglas, the islet itself is no more than a hectare, yet the litter at high tide level was awful. Jeff Miller asked one couple to record it. They noted forty-one thongs, fourteen light bulbs, seventeen 5 litre plastic cans of the sort that hold two-stroke motor oil, empty bottles of Gordon's Gin, Vat 69, Four Bells Rum and various Reckitt & Colman containers. The amount of pollution in the world's oceans is frightening. Plastic bags are a particular danger to the huge Leatherback Turtles, 1.5 metres across the shell. They feed on jellyfish and cannot distinguish them from plastic bags which catch on the spines at the back of their lower throats designed to hold back the bodies of the jellyfish while their stomachs squeeze the water out of them. When the spines are sealed with plastic, the squashed jellyfish escape with the water and the turtles starve. They seldom breed in Australia and they are probably now an endangered species.

We sailed on to the Arnold Islets north of Milman. These islets consist of a low sand cay separated by a wide reef from a higher wooded area. The water on the reef was less than knee-high when I walked across it. It is a living reef, a wonder of colour and texture: green, beige, mauve, pink, occasional circles of red, black, hard, corrugated, soft, wrinkled, honeycombed, feathery, with algae and some grasses growing there. Small rays, 35 centimetres across the wings, flashed through the water, parrot fish as bright as the birds that named them moved in schools. A small species of Giant Clam 20 to 35 centimetres long fed with their shells partly open. Algae grew thickly along the mantles of the shells, making them look like thick, corrugated, half-open, purple lips. As one approaches they slam shut and columns of water spurt in the air.

The tides were still wrong for recent turtle nesting but it was an altogether fascinating visit for me. After telling the story of the Macassan fishing for trepang for the Chinese market in my book *Sojourners* (University of Queensland Press, 1992), it was a wonder to hold the living creatures in my hands. The trepang there are an edible species but they are not the long, thick grey species from deeper water which make the highest prices when properly prepared. They were black, about 7 centimetres thick and up to 70 centimetres long. They anchor their tails in coral and stretch out along the sea floor like big worms. Five delicate branched antennae fan out in front of their mouths as they feed.

Five species of Mangroves grew in their chosen progression on the higher piece of land, one of them represented by half-a-dozen trees only. *Pemphis acidula*, a mangrove that needs fresh water, grew above the high tide level in a dense thicket 5 metres high. Sometimes these trees lose all their leaves in cyclones. Loose, dead coral piled high in long rows showed how much these islands are influenced by cyclones.

We saw our first turtle laying on an unnamed cay to the north of Arnold and almost level with the tip of Cape York. It was a grassed sand cay with one tree on it, a Beach Almond, *Guettarda speciosa*. It had an Osprey's nest in it (not then in use), as has some prominent feature on almost every island. These fine birds live on fish and are thriving in Australia. European egg collectors almost wiped out the northern species. In Scotland bird-lovers guard each nest twenty-four hours a day until the young are on the wing. Several Eastern Reef Egrets had nests in the tree, too, one with two bluish eggs. They use branches of dead coral as a foundation, since they cannot depend on driftwood, and they lay grass over it. The coral gives an angular, sculptured appearance to the nests.

There were a few tracks of Hawksbill Turtles that had come up to lay the previous night and one of a young crocodile. Male crocodiles from 1–3 metres long live about most islands within 100 kilometres of the shore. The breeding males allow young females to live and grow in rivers and estuaries, but they drive the males out to sea where they lead lonely lives until they have grown big enough to go back and challenge for a stretch of river.

We went back on to the island at night to see what was happening. Two Hawksbills were up on the beach. One had dug a couple of body pits and found the sand not to her liking, so she was making her way back to the sea. Jeff tagged her for identification with long-lasting numbered titanium tags. The other laid 133 eggs while we watched. Gulls and noddies cried overhead.

Very little was known about turtles until the late 1970s. The modern studies are urgent because turtles — their flesh, shell and eggs — are being harvested at a rate that no species can tolerate. Five species lay in Australia, the smaller Olive Ridley, the Loggerhead, Flatback, Hawksbill (which produces the thick, coloured scales that Japanese master craftsmen turn into superb combs and ornaments) and the Green, so good to eat and too greedily eaten. Some countries, ignorant of the life of turtles, resent criticism, harvesting what they claim as their own. Turtles travel the seas of the world, they are migrants on a grand scale.

Feeding grounds are usually distant from breeding grounds. When mating time approaches, males and females eat more than usual and build up body fat, then they set out on long journeys to the island on which they hatched or one near it. They eat little or nothing when travelling; some have been known to swim 2600 kilometres. It is a fine piece of direction finding. Turtles setting out from the rich feed beds at the bottom of the Gulf of Carpentaria have to swim north, then east, then south with currents and against cross currents and counter currents, as though they carry an exact map of the land and water, until they are near the chosen island which they locate by hearing, smell and sight.

They mate in the water, usually several hundred metres from shore. We had the chance to watch numbers of them. The males bite at the females, chewing at flippers, neck and head until they position them so that they can get a lift from a wave on to their backs and grasp them by throwing their foreflippers around the foreflippers of the females and hooking their long tails under their shells. Both males and females have cloacas, one orifice to satisfy all functions, but the males can evert their cloacas which open near the end of the tail to form a penis about 25 centimetres long. He slides it into the long, roomy cloaca of the female. Mating lasts for hours and he ejaculates a considerable quantity of semen but it seems a passionless union. The chief energy of the males is directed at maintaining their positions on slippery backs against the dislodging motion of waves. Most of the time two unmoving heads bob in the surf. Sometimes a second male levers himself with beak and flippers on top of a mating male. Since there is no orifice whatever available to him he floats joylessly and inconsequentially on top and three heads bob in the surf.

Sometimes when mating couples drift in too near the shore, breaking waves turn one or two males on to their backs. The water is not deep enough to allow them to flip themselves over and they wash helplessly ashore to die slow, painful deaths. We found one couple that had been rolled over. The same wave washed them both up the beach and they died side by side.

Over her week or so of sexual attractiveness, the female mates with several males. Their semen mixes together in her repository and she stores it until the first batch of eggs is ready for fertilisation when she measures out the right quantity of semen. By night she comes ashore to lay with the blind purpose of a bulldozer. Each species makes a peculiar track according to the length of the flippers and whether they use them in breaststroke or overarm fashion. If there is a nesting bird in the way it gets run over unless it is big enough and determined enough to peck at her eyes until she turns off course. Well above the level of high tide but before the top of the first sand dune, she begins to dig a body pit, flinging sand back and sideways with all four flippers. Sand gets on her back, on top of her head and in her eyes which stream salt water, freeing her body of salt and lubricating and cleansing her eyes, at the same time. When the pit is wide enough to give her plenty of room and about 45 centimetres deep she begins to dig the nesting cavity with her hind flippers, scooping the sand out with alternate movements. If she strikes rocks or roots, if the sand is too dry and runs back in, if it feels too hot or too cold (it must be 25° to 30°C), if something disturbs her by walking in front of her, she might rest and try somewhere else, she might go back to sea for a night or two. She is not compelled to lay until the hole feels exactly right.

When she begins to lay, she does not allow anything to interrupt her. She squeezes the eggs out at fairly measured intervals, usually one at a time, sometimes two or three together. Copious amounts of slime ease the passage. They pile one on the other, perhaps fifty, perhaps 150 – it varies with species.

When the last egg is laid she rests for a couple of minutes, then she fills the chamber with beautifully deliberate actions of her hind flippers. She drives them alternately into the sand at an angle, curves them into shallow scoops, then swings them over the hole. When sand reaches the top, she flattens her flippers and pats it down, adding sand until it is firmed level with the bottom of the body pit. Then, for some unknown reason, she continues to dig ahead through the sand for several body lengths, usually at right angles to her track up the beach before pulling herself out and working back to the sea. Probably it is to make it more difficult for a predator to detect where she has laid. In about a fortnight's time she will come back and lay another batch, then another and another. She might then go for a few years, several years, before laying again. A known Loggerhead turtle laid four clutches at the Mon Repos rookery near Bundaberg, Queensland in 1982. She returned nine years later and between November 1991 and January 1992 laid five clutches totalling 690 eggs.

Turtles can come up to lay in huge numbers. In 1984 on Raine Island in the far eastern edge of the Great Barrier Reef, 11 500 Green Turtles laid on one night along two and a half kilometres of beach. When such massive numbers lay, the latecomers sometimes dig up the eggs of the firstcomers.

The wild seas of an inopportune cyclone can wash out thousands of eggs. The Mangrove Monitor, *Varanus indicus*, seeks them out, but these big goannas seldom eat all the eggs in a nest. Dingoes, dogs and feral pigs take many eggs; in the 1960s foxes began to take the eggs laid on the mainland beaches near Bundaberg. Colin Limpus has watched foxes stationing themselves to wait for the finish of laying. A campaign of poisoning begun about five years ago has given the turtles another chance. But for years foxes ate 90 per cent of the eggs laid on northern beaches. Since the sex of the turtles is determined by temperature — warmth produces females, cooler conditions males (there is only about 2°C difference) — it would have been mostly potential females who were eaten. Mainland beaches are warmer than island beaches.

People are the principal predators. For Aborigines, Islanders, for the people of Papua New Guinea, for Indonesians and so many other races, turtles and their eggs are an accustomed food. When hunting was done by sailing craft or by rowing boat, the turtle population did not suffer. But increasingly they are now hunted in big, fast, engine-driven craft. Turtle products in Indonesia have become big business. Boats are travelling increasing distances to satisfy a growing market. Fishing trawlers throughout the world drown turtles of all ages by the thousand. A simple and inexpensive escape hatch can now be fitted to nets to stop this cruel waste. Its use requires education, not compulsion.

The eggs — the surviving eggs — take seven to twelve weeks to hatch. The young take a day or two to work their way to the surface. If they reach it during daylight, they wait until darkness makes the run to the sea a little safer. But one night on Mornington Island in the Gulf of Carpentaria Jeff Miller saw Water Pythons poised in striking position over the patches of sand where hatchlings would soon emerge.

The run to the sea is astonishingly quick and direct: there is no hesitation, no deviation. The passage of so many little bodies side by side leaves broad tracks of tiny scuffles in the sand as though made by animals 20 to 90 centimetres wide with hundreds of feet. On Lawson Island to the east of Cobourg Peninsula, Colin Limpus found that a crocodile had learnt what these marks meant. He had patrolled the beach and followed each trail to dig for unhatched eggs and hatchlings tangled in roots and grass, a common misadventure. Crabs take some on the run to water, Rufous Night Herons take a few more. They cannot swallow the hatchlings whole, they pick at them, tearing out the soft parts. The main predation takes place in the water. Fish, pelicans and young crocodiles collect to feast.

The hatchlings are thought to pick up a compass imprint during that first run which gives them a permanent alignment to the world. But initially it is the direction of waves that guide them. They swim into them and keep on swimming until they come to an ocean current where they hide in debris or a patch of weed, and drift. Their lives for the next fifteen years are unknown. On Bountiful Island in the Wellesley Group in the south of the Gulf, I picked up a year-old turtle which had been dropped by a predator, probably the Osprey flying overhead (the big White-breasted Sea Eagle father out would have been unlikely to drop it), only a few hours before. In all his years of turtle studies, it is the first time Colin Limpus has handled one of that age.

Tiger sharks and crocodiles feed on them while they are growing and when they are grown. All species take about forty years to mature; they live until they are about 100 years old.

We left Jeff Miller at Thursday Island and picked up John Winter, a consultant environmentalist of considerable repute. Thursday Island now looks too dry a little Australian town to support its lush history. An Indonesian fishing boat at anchor in the harbour provided the only exotic suggestion. John Winter's passion is possums and he wanted to refind Ringtail Possums that he saw on Vrilya Point on the western side of Cape York ten years ago. For some unknown reason, Brushtail possums are increasing in Tasmania to the point of being a nuisance while, for an equally unknown reason, both Ringtails and Brushtails are decreasing alarmingly on the mainland.

We travelled far to the west of Thursday Island to distance ourselves from the wide line of formidable reefs to the south, then we played the part of a sailing ship of 150 years ago and "worked to weather", beating south against a south-east wind force four to five (15 to 20 knots). We travelled south of Vrilya Point, then worked easterly towards it, "tacking in a funnel", as the navigator explained. That was more explicit than the triangle he had drawn on his chart. It allowed for the height of our sails and gave us a three-dimensional passage. But we had to use the engine eventually or we would have lost a whole day.

For most of the night, three of us searched eucalyptus forest, vine scrub and melaleuca swamp with torches and spotlights and we found no possums, though that need not mean that they are gone. There were never many there. We did see a White-tailed Rat. I had never seen one before since they are confined to the tip of Cape York. Our biggest rodent, it watched us quietly for several minutes with its spectacular 34 centimetre tail hanging down the other side of the low branch it sat on.

John Winter left us at Weipa; Colin Limpus joined us. Before he began his turtle research, he spent ten years studying freshwater crocodiles and tells stories of racing across rocks by torchlight to stop crocodiles that he wanted to tag escaping down rapids to the seclusion of deep water. Wildlife scientists do not spend staid lives behind a microscope. His twelve years of work with turtles have gained him worldwide recognition.

On the long run south over a confused sea to the Wellesley Islands, everything on the water interested Colin. "That fellow is writing down every bird and snake and turtle he sees," said one of the crew. "He must be out of his mind."

"No, that is his mind," said somebody else.

In an hour's run, with the help of all the passengers, he counted about 100 seasnakes of four species. We must have followed a line where two ocean currents met, bringing together a rich supply of food. It is the greatest number of seasnakes ever recorded in a small area in Australia. Little is known of these extraordinary creatures. Most of them are strikingly coloured and their tails have flattened into broad paddles. A couple of years ago when diving on the Chesterfield Reef, 500 kilometres off the east coast of Australia, my daughter found hundreds of at least five species living around a high, underwater, rocky pinnacle. The snakes were curious and insisted on inspecting the divers, so they stayed still while snakes swam between their legs and bumped inquiringly against their facemasks.

We stopped first at Bountiful Island, named by Flinders for the number of its turtles in 1802. As he wrote in *A Voyage to Terra Australis*, "We contrived to stow away forty-six, the least of them weighing 250 lb, and the average about 300 [136 kilograms]; besides which many were re-turned on shore, and suffered to go away."

It was a common thing to stow turtles on the old sailing ships. They placed them on their sides in the bows where they leant together like books that have slipped sideways on a shelf. With buckets of seawater thrown over them to keep them cool and moist, they lived for months, an excellent source of fresh food. The sailors collected eggs of seabirds, too, extravagantly breaking however many thousands they first found on an island so that they could collect fresh eggs.

Bountiful Island is well vegetated. There is a great variety of plants on it. However, it is not lush. From a distance, even at close quarters, it looks dry. But it attracts turtles. We caught more than sixty Greens on the first beach, rushing into the water to roll them on their backs then towing them up the beach to be measured and tagged. It is hard work turning over 140 kilograms of energetic turtle and it hurts to get a hand slammed against a shell with a blow from a flipper, especially if its edge has a few barnacles growing on it. One does not put a hand near the mouth. It can crush fingers.

Our hard work was not really necessary. There were easier turtles to find. Farther round the island, in a shallow rock pool, 166 basked side by side in the sun, all of them sound asleep with enough water under them to keep them wet. "Stranded turtles" is the scientific term for such congregations. It is a silly term. These were adult turtles with perhaps sixty years' experience of tides. I believe that they knew exactly when the tide would come in and wake them up.

On the next beach, swimming, basking, mating, I counted 275, necessarily a rough count because numbers of them were swimming in deep water.

A few Green Turtles had come in to lay the night before, a few Flatback nests had hatched. Colin dug up a young Flatback that had got caught in grass roots. The hatchlings of this species are unmistakable, they look startlingly new. The greyish-green scales are outlined in black as though they have been recently painted

A small number of Caspian Terns nested on a sandy ridge, beautiful big birds with strong red beaks. They resent intrusion and fly noisily overhead when one approaches. Gulls steal eggs whenever opportunity offers. This is not too serious as terns simply replace missing eggs. But on exposed sand, it is the function of the brooding female to shade eggs during the day to keep them cool and to warm them at night. On a hot day, even twenty minutes' exposure can addle the eggs, and the female, not knowing anything is wrong, sits uselessly for the rest of the time.

Rocky Island, close by, supports two species of plants only. It has been taken over by Brown Boobies for nesting. The island was white with chickens standing on a deep white paint of excrement. Unknown rats (we chased down three for indentification) and Pelagic Geckos forage at night in thousands.

We ran up towards Pisonia Island but it was not safe to approach it. These seas are more or less uncharted. Mitch put a lookout high on the mast. "What can I say he is standing on?" I asked Sally Anderson, who at nineteen has her master's certificate. "He's standing on the spreader on the futtock shroud," said Sally. But the sea had risen too much for any lookout to see rocks. We turned back to safer water.

Greg Leach, a botanist with the Conservation Commission of the Northern Territory, joined us at Nhulunbuy and gave us a new dimension, one essential for my research for a new book, *The Growth of Australia*, since I need to know the progression of plants that stabilise, then cover, islands. We crawled through mangroves, tore a way through vine scrubs, scrambled over a dead tangle of cyclone debris now masked by young growth. And Greg always had eyes for plants, whether at his feet or 30 metres up. He was looking especially for exotic weeds and, to the delight of all of us, he did not find any.

We visited island after island along the north of Australia, we made a run of several hours with every sail unfurled and holding wind. Even the mainsail spread without its usual reef. The sea was flat, the wind light, and David Nash, first mate (he had helped build the ship), was showing her full glory.

On Truant Island, up near the Arafura Sea, we found a series of wells dug by Macassan trepang fishermen last century. A big clam shell that had been used as a baler lay half buried in silt, a couple of old Tamarind posts still marked where a guard rail had been. We found a Coral Tree, not known on the Northern Territory mainland, about to break into bloom. We found the tracks on a beach of an egg-laying seasnake, *Laticauda genus*, never known to breed in Australia. We found islands where significant numbers of Olive Ridley turtles lay and we found their hatchlings which had never been seen here before. We saw Black-naped Terns courting each other with delicate offerings of little silver fish.

What we saw on these distinct, different, distant islands was the infinitely slow process of creation.

The blue-green Darling River

In November 1991 the great Darling River in western New South Wales painted itself a bright poisonous green. A thick ribbon of blue-green algae wound a thousand kilometres from Mungindi to Wilcannia and lay there, sometimes on top of the water, sometimes submerged, until a lucky December flood washed it away. It was the greatest algal bloom ever recorded on any river in the world. What caused it?

Such a dramatic announcement was needed to draw attention not only to the Murray–Darling basin, but to the Hawkesbury River, the Nepean, to every dam and watercourse in New South Wales. The water in most storage dams has the capacity to kill, the fish in them might or might not be poisonous.

In September 1992, Elaine and I began talking in Sydney about blue-green algae. We circled out through Orange, Dubbo, Nyngan, Wilcannia, up the Darling River through Tilpa and Louth to Bourke, then back to Sydney along the Barwon River via Brewarrina and Walgett. We read every word we could find on the subject, including some official reports too sensitive to be released. We listened to scientific fact, to prejudice, to old wives' tales, to Aboriginal dreamtime stories, to fascinating observations, to passionate descriptions of work being done, to condemnation of work being done, to expositions of what ought to be done, to wild exaggeration, to misconceptions.

Everywhere we heard the cotton farmers blamed. "There was no blue-green algae until they started to grow cotton", which is partly true, but the cotton farmers are not the prime cause, they are merely the last straw, the final insult to a long-abused river. Here is the complete complicated story of blue-green algae.

Algae make a division of plants which do not differentiate flowers, leaves, stems and roots. They are shapes synthesising chlorophyll, some minute, some as big as trees, many of them exquisitely beautiful. The lichens of Tasmania's cool forests are wondrously delicate compilations of fungi and algae. They grow under light so curtained by leaves that one needs a torch to see them clearly, a microscope to define their complication. For what eyes are they made?

Our rivers, creeks, lakes and billabongs, all our waters, support hundreds of species of green algae. They are an essential food for an astonishing range of life: for viruses, bacteria, for the hosts of creatures invisible and barely visible known by the unfamiliar names of zooplankton, protozoa, rotifers, microcrustacea, for shrimps, yabbies, fish, even for some species of duck like the remarkable Freckled Duck which might be a living relic of the ancient origin of geese, swans and ducks, or the spectacularly striped Pink-eared Duck. Its bill is fitted with a gill-like fringe so fine that it strains microscopic plants and animals out as the bill opens and closes. These green algae vary in size from 0.015 millimetres to 1.5 millimetres. There are leech-like shapes, rods, crescents, hairy ovals, waisted ovals, spheres or lumps made up of a congregation of minute cells, eight-rayed stars, many-pronged blobs like particularly vicious double cat-heads, regular and irregular shapes with single and double filamental tails that drive them through the water as though they are animals, cells that link together in the form of chains or threads millimetres or centimetres long, or shapes as perfect as crystals patterned with mystic designs.

Blue-green algae have recently been divorced from true algae. They were difficult to classify since they do not have a nucleus, the central repository of genes, which is an atttribute of bacteria, but they synthesise chlorophyll, a function of plants. So they have been given a status of their own, cyanobacteria (blue-green or photosynthesising bacteria), though scientists sensibly decided to retain their older name. Despite their late classification, they are of immense age, 3000 million years, which makes them one of the earliest forms of life. They bloomed in that cloudy beginning far more massively than they do now and during millions of years they created the world's oxygen. They are still an essential part of life in water, producing nitrogen for those plants that need it, an ability which makes them particularly valuable in the unnatural environment of a rice field.

There are many genera, many more species, but only about a dozen of them cause trouble when they reproduce in unnatural numbers. The marine specimens that cause the frightening red tides are among them. They mass in long broad bands like huge spills of red paint, killing fish and molluscs. There are periodic outbreaks on the Great Barrier Reef and in Western Australia; the American coasts are particularly troubled. Some edible bivalves there seem to be unharmed by its poison, but they secrete it and poison anything that eats them. The popular clam can be deadly to humans. Mussels grown in the big Peel Inlet south of Perth were recently found to be poisonous.

The freshwater species that took over the Darling River were *Anabaena circinalis, Anabaena spiroides* and *Microcystis aeruginosa*. Although a bloom looks like a shapeless, stinking green slime, a microscope that magnifies about 700 times only reveals fascinating units of cells. The roughly spherical (coccoid) cells of *Microcystis* collect together in irregular clumps about one-fifth of a millimetre wide; the cells of *Anabaena*, measuring fifty to a millimetre, join together in flexible threads up to 3 millimetres long. Each filament of about ten to 150 or so cells acts as a single entity with one or two special nitrogen-fixing cells in it and usually an akinete, a resting spore of considerable durability. They can survive in water, in mud, in ice, in dust for up to eighty years then come to life. Wind and water disperse them, so do the feet of waterbirds and other animals.

Being bacteria, blue-green algae increase by cell division. The conditions which allow unrestrained generation, a bloom, are clear, still, warm water (20 to 30°C), such as that held back by the weirs of the Darling River when there is no flow, abundant zooplankton (minute animal life) which eat competing green algae and avoid the poisonous blue-green, high levels of phosphorus, and alkaline pH levels of 8 to 10. The number of cells in a bloom are infinite, some Darling water yielded counts of 600 thousand to a millilitre, 3 million to a teaspoonful. Fifteen thousand cells to a millilitre is a supposedly safe limit; smell and taste become decidedly nasty at 2000 cells a millilitre. In July 1969 the Sydney *Sun* reported that water from the Prospect Reservoir tasted "like a billabong with the swagman still in it".

Like all successful organisms, blue-green algae has extraordinary abilities. Each cell is equipped with a vacuole, an internal bladder which it can inflate or deflate quickly or slowly according to how it wishes to move. If they are feeding and floating with full air sacs on a cloudy day, a sudden burst of bright sunlight sends them plummeting to the bottom of the river. In changing weather the whole bloom bobs up and down like billions of billions of Cartesian divers linked together.

Their behaviour is governed by complicated chemical processes. Gas vacuoles collapse, for example, when suddenly squeezed by cells swelling and stiffening by light-triggered uptake of potassium ions with a simultaneous production of carbohydrates. But the cells do seem to have some form of immediate communication — it must be chemical — though their whole behaviour suggests that they are not only one of the original forms of life but one of the original forms of intelligence.

In big dams they have been collected at depths of 50 metres. The richest strata of nutrients varies in dams and rivers according to temperature, light, current or any change in conditions whatever. By manipulating the air in the vacuoles, blue-green algae drift up and down seeking out the best feeding ground. They take advantage of excess food by storing what one scientist called "a luxury amount of reserve phosphorus" to tide them over leaner times.

A great excess naturally triggers propagation. When coming to the surface to divide, they bring extra food with them as an immediate supply for the new cells. A dying cell releases its phosphorus and neighbouring cells absorb it. Sometimes another species works its way into the mass. They often exist together; sometimes the first species dies and the other takes its place.

The poison that algae produce to protect themselves from predators is variable and unpredictable. Three samples taken 40 centimetres apart in a bloom of one species might be harmless, mildly toxic and deadly. At its worst it is ten to fifty times more poisonous than strychnine and acts just as quickly. The poison withstands freezing and boiling. Since some true bacteria live in harmony with blue-green algae, there is always a danger of other toxic species generating. Conditions under a thick dying bloom seem to be ideal for the awful bacteria which cause botulism. It is 300 thousand times more poisonous than the most deadly blue-green algae.

In the 1970s, Valerie May, a phycologist who made a special study of toxic algae, watched several turkeys come down to a farm ground tank at Condobolin in central-western New South Wales. They began to peck at clumps of *Microcystis* on the edge of the water. Within minutes they all dropped dead. The action of the poison differs with species and with the astonishingly variable degree of virulence. The most potent cause nerve and muscle failure and a quick death due to respiratory arrest. All cause liver damage which can lead to a slow death when insidious low doses are absorbed. In an experiment in 1975 White Leghorn Chickens injected with repeated low doses for months remained apparently healthy but on examination their livers showed marked degeneration. People who swim in affected water seem to suffer highly increased incidence of eye and ear troubles, rashes, asthma, hay fever and diarrhoea. But these are all too-common summer complaints and cannot yet be definitely attributed to blue-green algae. In 1979, 149 very sick Aborigines, most of them children, were admitted to hospital after swimming in a dam on Palm Island off the coast of north Queensland. The cause was probably blue-green algae, but no proper investigation was ever made. The only reliable test yet for its poison is physical, injection through the peritoneum into the abdominal cavity of mice, and it is not really satisfactory since it can take up to a week to get final results. Naturally, high virulence shows up in minutes with the death of the mice, but low virulence involves examination of the livers after several days. A method of chemical analysis in laboratories is being developed to detect the still-dangerous low levels of toxicity.

The first report of stock deaths in Australia came from Lake Alexandrina at the mouth of the Murray River in South Australia. In 1878 numbers of cattle died when they drank the water blooming with blue-green *Nostoc spumigena*, a species that has not shown up in Darling River samples. Over the years some thousands of sheep and cattle have been poisoned. No doubt many of the individual deaths attributed conveniently and carelessly to snake bite were caused by blue-green algae. It is often difficult to be certain about the death of numbers of animals since the worst cases occur during droughts when conditions of warm, slack water so favour the blooms and sheep and cattle are dying anyway. At the end of 1991 Ted Davies of Murtee Station near Wilcannia lost 500 sheep from what he feels certain was blue-green algae poisoning; a veterinary surgeon agrees with him. They were watering at Lake Oxford on another property he owns 20 kilometres east of Ivanhoe. There was not a massive bloom on the lake, but a constant wind from one direction over several days blew all that there was to the only place where his sheep could water. When he found them in trouble they were labouring for breath and gasping in pain. He could hear them from many metres away.

Sometimes sheep are poisoned when cattle drinking the same water escape, since cattle have the length of leg to walk out to clean water beyond the green slime along the bank. The poison is not in the water, it is in the cells, so animals cannot get poisoned unless they swallow them. The poison is for defence, the cells release it when they are threatened. Sheep and cattle do not like the smell of blue-green algae any more than humans do, and will walk many kilometres in search of clean water, although sometimes starving stock will deliberately eat it, wading into the water and scooping up the slime with bottom jaws. Ducks seem to be immune to the poison, so are most fish, but when feeding on blue-green algae their flesh can be poisonous. No one knows how poisonous it is, or how long it lasts. Certainly the livers of both ducks and fish could be deadly.

The first news of the blue-green Darling came to Allan Amos of New South Wales Water Resources at Bourke on the night of 30 October 1991. Garry Mooring of Rose Isle station between Bourke and Louth rang him to ask had the river gone rotten or had someone dumped a chemical in it. Garry had been out on a motorcycle checking his drought-stricken sheep. He was about a kilometre off the river when the wind carried an awful stench to him, a nauseating fetid earthiness, much like the smell of drains for pig manure in intensive sheds. He quickly traced it to a strange slime on top of the water and rode home for his boat to try to find out where it was coming from. In places the scum was so heavy that the boat seemed to be pushing against solid lumps. He put an arm into the water and the slime coated it like green paint. He had to find a clear patch of water to wash it off.

Allan Amos knew at once what it was. For many months he had been testing water for salt at intervals along the river and in May he had found an abnormal amount of blue-green algae in a sample he took near Brewarrina.

By the first week in November the whole river was affected. Billions of bacterial cells dividing and dividing out of control can soon reach infinite numbers. Brian Haisman of the Department of Water Resources at Dubbo was in the helicopter that made the first investigation of the river. He found it a stunning experience, no one had believed it could be as bad as it was. When the first television pictures were broadcast, most people who had known the river all their lives simply shrugged and said "They must have faked it".

The reality was a considerably distressing experience, especially for Aborigines and frontage landowners who are so intimately connected with the river. This wondrous living thing hundreds of thousands of years old seemed suddenly to have died. It was to be avoided like a poisoned carcass.

Aborigines believe that Baiame himself, the great creator, made the Darling River. He strode the length of it and his feet pressed down the channel, while his two Spirit Dogs gambolled about and hunted and created the tributaries. Later he showed the original people how to make the great stone fish traps at Brewarrina and for thousands of years all the north-west tribes, at least ten, met there each year for a week of feasting and ceremony. The river supplied a great deal of a year's food and water for all the tribes on it — until November 1991 it still did. Aboriginal people prefer to drink river water; they regard it as their water. When we spoke to Essie Coffey at Brewarrina — she is one of the leading Aboriginal women in Australia — she took us down to the river bank near the beginning of what is left of the fish traps. She would not consider coming to lunch with us to talk. "I can only think on my own grounds," she said. "My bible is the sacred sites, my faith is my people."

Europeans on the river are living on stations named 150 years ago by the first adventurous squatters: *Tapio, Burtundy, Moorara, Tintinallogy, Murtee, Curranyalpa, Gundabooka, Yanda, Jandra, Mogil Mogil, Collymongle,* numbers more. For sixty vital years the river carried the paddle steamers bringing station and homestead supplies from stud rams to silk ball gowns, taking back the wool, skins, live sheep that paid for them. Periodically, the river irrigated thousands of hectares of their runs, germinating months of superb feed and never coming down in the wild, destructive floods of its upper tributaties or the undisciplined coastal rivers. It gave days, weeks of warning that it was on its way with a wave of rich water 5–20 kilometres wide which would creep over land expecting it. The river was a prized possession. I lived on a river, the Namoi, for twenty years. One pays a great deal more for riverfront properties and although ownership rightly stops at the bank, one naturally has a proprietorial attitude to the stretch within the boundary fences.

Overnight the river changed from an asset to a disaster. Was it safe to leave stock on it, to water the vegetable garden from the domestic electric pump on it, to fill the swimming pool for the children by the same pump? What of the bigger pump driving out stock water to the back paddock 10 kilometres away? Was it filling the tanks and troughs with poison? Was the water safe to wash in? Some wondered what they were going to drink, the drought had gone on so long they were out of rainwater. Others even wondered what was going to happen to wildlife. Was everything that depended on the river going to die?

The Darling was a bitter disappointment to Charles Sturt who followed the Bogan River on to it at the end of 1828 in what is probably the worst drought ever known in that country. Aborigines were starving, some were dying. Sturt needed water for his men, his horses and his working bullocks. At first the river was a magnificent sight: "We suddenly found ourselves on the banks of a noble river," he wrote in *Two Expeditions into the Interior of Southern Australia*. "The channel of the river was from seventy to eighty yards broad, and enclosed an unbroken sheet of water ... literally covered with pelicans and other wild fowl ... the men eagerly descended to quench their thirst, which a powerful sun had contributed to increase; nor shall I ever forget the cry of amazement that followed their doing so, or the looks of terror and disappointment with which they called out to inform me that the water was so salt as to be unfit to drink! ... on a closer examination, the river appeared to me much below its ordinary level, and its current was scarcely perceptible."

They found fortuitous fresh water in a spring-fed pool off the river on the other side (no such springs exist now) and they continued downriver for about 90 kilometres. The river was salt all the way, and Sturt marked his map with three brine springs "from which a considerable stream was gushing". Upstream where the Castlereagh joined the Darling the water was even saltier. It did not trouble the fish; he found "deep reaches" alive with them.

When Thomas Mitchell explored the Darling in 1835 there was a strong flow of fresh water in the river, yet he too found a salty pool a little way downstream of the junction with Gundabooka Lagoon (now Humes Creek) which was Sturt's southernmost point.

The blue-green algae story can only be understood if one knows how the Murray–Darling basin functioned before Europeans so grossly interfered with it. Whoever designed it did a truly brilliant job. The most remarkable feature of the rivers was their inbuilt filters. No muddy water with its rich nutrients ever flooded out to sea. It was either strained through amazing complexes of weeds or trapped in ephemeral lakes and billabongs to settle out its precious deposits. Pre-European Australia could not afford to waste any phosphorus. Big creeks like Baradine Creek, which flows north into the Namoi River, did not have any defined channel into the river. For the last 20 kilometres it spread its muddy water over grassland and the river received slow, clear water. The Paroo River which came down from Queensland in wide floods usually never reached the Darling. It spread out into lagoons and lignum swamps. The upper Lachlan River filtered through a series of marshes; before it met the Murrumbidgee it passed through the Great Cumbung Swamp 100 kilometres long by 25 kilometres wide. The Murrumbidgee cleared itself before it met the Lachlan through what were described as "vast fields of Polygonum on both sides of the river". They were 140 kilometres long, a dense tangle of the two-metre high shrub now known as lignum. One big area on the south bank called Lowbidgee has been declared "sensitive land", the only such classification in the state. Neither trees nor lignum can be cleared. Lignum followed down the floodways and overflows of the Darling, too. Mitchell reported it 400 to 1600 metres wide.

One of the main filters of the basin was the Macquarie Marshes, once about 100 thousand hectares of the 3 metre tall reed *Phragmites australis*. Oxley was confused by them in 1818, but a huge flood gave him no time to explore them; he had to move east to high ground. Sturt made his way through them and around them with considerable difficulty: "We found it unsufferably hot and suffocating in the reeds," he wrote, "and were tormented by myriads of mosquitoes, but the waters were perfectly sweet to the taste, nor did the slightest smell as of stagnation proceed from them." The filtering job was perfect.

I have put all the descriptions of these marshes in the past tense. One of the Darling's main troubles — it is probably its chief trouble — is that most of these filters and settling ponds have gone, none of them are as effective as they used to be. All are stocked, some are farmed, water has been diverted around some of them. They once had the capacity to deal with even the absurd amounts of phosphorus (the cause of the blue-green algal bloom) that we are putting into our rivers.

An astonishing diversity of plants made up the filters. Of the tall plants like lignum and reeds already mentioned, apart from River Red

Gum and Coolibah, there were also cumbungi, canegrass, manna grass, Giant Sedge, Giant Rush up to 4.5 metres high, and dozens of lesser plants under the popular names of sedge, rush, nutgrass, water-milfoil, waterwort, mudmat, mudwort, hornwort, knotweed, spike-rush, Water Primrose, Dirty Dora, all of them equipped by density or leaf surface to slow down water and collect its solids, even in solution. In the rivers themselves grew several species each of waternymph, arrowgrass, eel-weed, starfruit, swamp lily, duckweed, pondweed with floating leafy stems up to 4 metres long. On our 2-kilometre section of the Namoi River at Boggabri that I knew so well for twenty years, Couch Grass, *Cynodon dactylon*, climbed down the bank into the river, Water Couch, *Paspalum paspalodes*, climbed out of the river up the bank, wide rafts of Red Azolla, a pretty floating river fern, lined the water's edge and several other plants grew in deeper water. Azolla has an important symbiotic relationship with blue-green algae, one of the species of *Anabaena*, which fixes the nitrogen it needs like the nodule-producing bacteria on the roots of legumes.

All these plants supported a busy complex animal life, a huge associated world. A small snail fed on Azolla in millions, one could bring up twenty in a single handful of fern and water. At times of low water, Freshwater Catfish and Silver Perch fed on them in such numbers that their flesh became tainted with a strong muddy flavour. Now the Azolla is gone, the snails are gone, the catfish and Silver Perch are almost gone. (Of 10 thousand fish caught recently at Bourke Weir, only seven were the once common Silver Perch.) All the plants that grew profusely in all the northern waters of the Murray Darling basin only fifteen years ago are now almost gone and with them their associated life. Some hundreds of species of plants and animals that enlivened our streams have been suddenly reduced to insignificant and useless numbers. Wondrously busy rivers are now barren irrigation channels. It would not be exaggerating to call it the most serious malfunction of the Australian environment. Our rivers cannot live without the plants, we cannot live without our rivers.

And nobody can do any more than hazard guesses as to what has happened to the plants, even Geoffrey Sainty and Surrey Jacobs, the authorities on native waterplants. They wrote the waterplant chapter in *The Murray*, a superb work produced by the Murray Darling Basin Commission in 1990. There is a decline in plants in the Murray and immediate tributary streams, but it is not nearly as serious as in the upper Darling basin. Introduced European Carp are widely blamed. Brought into Victoria illegally as a food fish about 1960, stupid Gippsland farmers stocked dams big and small with them, then unknown, irresponsible cotton farmers transported them to irrigation channels as weed eaters at Wee Waa on the Namoi and at Narromine on the Macquarie. In the great flood of 1976 they populated the entire Murray–Darling basin. Certainly they were in great numbers for a few years. In two days fishing in the Namoi River at Pilliga, my wife and I caught more than 200 between 2 and 12 kilograms (a tonne or more) and eight native fish only (good Golden Perch). European Carp are eaters of weed and in search of other food they fan the muddy beds of rivers and lagoons with their fins, disturbing roots and reducing necessary light to the plants by muddying the water. But the water weeds began disappearing from the middle Namoi before carp were abundant and they continued to disappear at a much faster rate in all waters during the late 1980s when European Carp had been greatly reduced. Golden Perch and Murray Cod found their eggs and fingerlings good to eat and are now thriving again and apparently dominating.

I believe that European Carp had little to do with the disappearance of the weeds. (It would still be good to get rid of the carp. Selective electronic stunning offers hope of lessening them. They could be marketed as pet food, even as fertiliser. Although those in faster coastal streams can be very good eating, those from muddy western rivers are usually foul.) Five other factors are making the watercourses unfavourable to plant growth: the cold water released from deep dams for irrigation; the unnaturally high level of the summer flows of this cold water; increased turbidity due to erosion of both farm and grazing land; increasing salinity; and increasing nutrient levels, mostly caused by phosphorus. The once little-known word "eutrophication" is now in common usage. Native water plants probably dislike phosphorus as much as heath plants but it is difficult to believe that their extinction has been caused by a gradual build-up of nutrients suddenly reaching an intolerable level. Increasing salinity has changed the dominant plant of the Great Cumbung Swamp on the Lachlan from Cumbungi to *Phragmites*, but there has not yet been any change to the salinity of the Darling. The only sudden change to the whole system is the increasing use of chemical sprays on cotton, corn and other crops. Perhaps these plants are super-sensitive to either the hormone weedicides used in all farming or the defoliants used to dry cotton and corn before harvesting. It is imperative that scientists be appointed to retest every spray now in use, especially the insecticide Endosulfin used on cotton. Tests are done on every spray before it is released — that is a requirement of law. But now they are not nearly comprehensive enough, and penalties for the misuse of sprays are trivial. Towns get sprayed too often and also farmhouse roofs that catch rainwater; careless pilots do not turn off the spray as they swing over a river to make the next run. There is no evidence about how dangerous Endosulfin and other sprays are. One cannot produce human corpses a few hours after spraying, so there is no official concern. Someone's cancer five years later, several unthrifty babies in twelve months' time, a few sick dogs, even a spate of miscarriages in the ensuing weeks can be ascribed to other causes and sometimes, indeed, should be ascribed to other causes.

But, as too frequent overruns on to the Gwydir River have shown, Endosulfin is a fish killer. Each time there have been the immediate bodies to prove that, tonnes of them, thousands of fish. Has this spray and others like it, now used somewhere on every tributary of the Darling, caused the death of the multitude of animals visible and invisible that depended on the waterweeds and did their death cause the death of the weed? Is there such an intimate association between plant and animal life, including the many viruses and bacteria, that neither can exist without the other?

These chemicals can have an insidious effect in sub-lethal quantities. Work in 1988 by V.J. Pettigrove on the larvae of midges, *Procladius paludicola*, taken from the Murray and Darling Rivers showed deformed ligula (mouthparts). Normally hand-shaped, some mouths had lost all the finger-like extensions — they were left with mere blobs — others had four fingers, some nine.

The first attempted solution to the Darling bloom came from the Bourke Shire. At the direction of its president, Wally Mitchell, it sprayed 60 kilometres of the river, banked behind the 20A weir near Louth, with Coptrol, a compound of copper. "I put a window in the Darling," says Wally. "We had clear water." Others saw a rather smudged pane with too many cracks in it. Sue Salmon of the Australian Conservation Foundation over-reacted and on 12 November 1991 issued an inflammatory press release headed "Shire takes law into its hands to poison Darling River". The manufacturers of Coptrol threatened to sue so she had to withdraw most of her statements the next day. Such antics do a great deal of harm to these necessary watchdog bodies. Too often they present reaction instead of reason and lose their credibility.

Copper in some form, usually copper sulphate, has been the standard method of treating outbreaks of blue-green algae. Adelaide has used it regularly on its water supply dams for many years. It kills fish and frogs (and what else?), it leaves a deposit of heavy metal on the bed of the storage and it has to be used very early because a developed bloom can release so much poison as it dies that the water would be unsafe to drink for weeks. In a dam a deposit of metal on the bottom quickly gets covered with silt and rendered harmless. A river bed gets stirred up repeatedly. There is a further danger with the use of any compound of copper in rivers. The chemistry of the water varies with the soil on the banks: it can be acid in one place, alkaline in another. Acidic water dissolves poisonous copper salts before they settle to the bottom.

It is a good thing that the Darling River made such a startling display of its predicament. Valerie May, who did the early Australian work on blue-green algae, told me that about twenty-five years ago she submitted a paper on its danger to a scientific journal. Its referee refused her on the grounds that blue-green algae had never been a major problem and never would be. In the late 1920s, it had made a spectacular bloom on the huge Hume Dam above Albury as it filled with new Murray River water rich with disturbed earth. It took 22 tonnes of bluestone (copper sulphate) to subdue it. Farm dams and ground tanks bloomed regularly, but few realised that river water locked behind weirs presented the same conditions.

Very soon after the public announcement by the Darling River, Water Resources formed the Blue-Green Algae Task Force. On 22 September, after nine months of intensive study, it released an excellent report. The fourteen scientists who composed the force came from thirteen organisations such as Water Resources, New South Wales Fisheries, Hunter Water Corporation, University of Technology Sydney, Conservation and Land Management. It was a rare case of bodies which are often strangely antagonistic to one another working together. At the launch of the report at Parliament House, Ian Causley, Minister for Natural Resources, usurped the credit for the initiation: "I established the Blue-Green Algae Task Force ... I gave the Task Force nine months to complete the job."

It will take ten to fifteen years of intensive care to cure the Darling, but fortunately there is immediate effective treatment for the symptoms and a number of ways to dodge the poison. Some of the cures will meet murderous opposition. I do not use that adjective lightly. One engineer who advocated what I regard as very mild changes to irrigation methods has already been beaten up, an official in a high position told me that he has heard whispers of concrete shoes waiting anyone who interfered with water rights. There are millions of dollars involved in land preparation for cotton and insufficient water to satisfy the licensees. In February 1992 as the December fresh was washing the lower Darling clean again, Water Resources issued an "Interim Unregulated Flow Management Plan for the North-West". Unregulated water is that spilling from full dams or flowing from flooded uncontrolled creeks. This water has always been regarded by irrigators as free water. It is what they pump into huge earth-banked storage tanks to water extra unlicensed ground and thus they even out all flows down the river. Any water that escapes their pumps has been regarded as waste water. In an essential move to reimpose some sort of natural flushing to our waterways, Water Resources has given the right of first choice of this water to the Darling River. Between October and April each year at least two flows of 2000 megalitres a day have to be allowed to pass freely down the Darling all the way from whatever tributary is in flood to Wilcannia. Furthermore, during the period September to February, there have to be two bigger flows of five days each (14 thousand megalitres at Brewarrina and 10 thousand megalitres at Bourke) to allow fish to migrate over the weirs. Seventeen weirs on the Darling have stopped the natural movement of native fish, once accustomed to travelling the full extent of our rivers, journeys of 1000 kilometres and more.

There has always been a great deal of illegal pumping. "When the lights go off, the pumps go on" is a Bourke maxim. To overcome this, big irrigators had to fit Time Event Meters on their pumps by October this year. They record how much is pumped and at what time. Smaller irrigators must instal the meters by October next year.

There is one disturbing sentence in this draft plan: "The large scale use of stored water supplies to achieve any worthwhile changes in the Barwon–Darling would result in serious economic and social effects."

And so it certainly would — cotton is New South Wales's second highest earner of export income. But the Murray–Darling basin provides a quarter of the farm gate value of Australia's agricultural production. The entire basin has to be considered; all that long wide strip spreading down from Queensland through New South Wales into Victoria and South Australia is under threat.

But there has already been bitter oppositon to the interim plan. Although Queensland has joined a body known as the Murray–Darling Basin Initiative and joins in all discussions, she simply does not understand any water allocation for environmental reasons. Even New South Wales landowners on the Darling and Barwon Rivers have sharply different ideas. Brian Haisman with Water Resources at Dubbo told me that it is fascinating to hear people from Wilcannia talking to people from Collarenebri: "It's like a meeting of people from different planets." Ideas will have to mature quietly, he said, nothing can be rushed. The whole river population is passing through "a tempestuous adolescence".

Another quick move by Water Resources was the sinking of ten bores for stock water 50 kilometres apart on both sides of the river from Wilcannia to Bourke. Unfortunately the move was too quick for the hydrologists and some of the bores missed the best water. But they did prove that it is possible to tap sand beds (probably ancient rivers) in many places that yield quantities of good water at shallow depths, 18 thousand litres an hour at 30 to 40 metres. These bores, lined with polythene or polyvinyl chloride and fitted with electric pumps, can be put down for less than $6000. Increasingly landowners will depend more on underground water than river water.

Some shire engineers are considering aeration of the water about the intake of pumps filling town reservoirs. Big air compressors will not only provide a protective curtain of air to keep blue-green algae at a safe distance but also their agitation of the water will discourage the blue-green algae. It thrives only in still water. Activated carbon filter plants are another source of safe water and some towns will instal them, although they are messy and expensive to operate. A plant was already ordered for Walgett before the massive blue-green algae outbreak. For the last twelve years the townspeople had been noticing an increasingly nasty taste in the town water and they avoided swimming in the river in the hottest months — it smelt too much of dead fish and pig manure, the typical blue-green algae smell.

We had a look at a plant which has been operating for years at the Gingie Aboriginal Settlement out of Walgett. Wally Jones, the operator, keeps about 50 thousand litres of treated water in supply tanks for the small community. Working the plant is a full-time job. An electric pump pulls the water out of the river into a cone-shaped flocculating tank fitted with circular baffles where it is mixed with alum to precipitate solids, then the water, hardened by the alum, flows down by gravity through sand and soda ash to soften it again. In the final step the water is forced through activated carbon to strip out the chemicals and any poison from blue-green algae. This expensive and excellent filtering medium looks like soot. It is granulated charcoal heated to about 900° in a current of steam to remove hydrocarbons. Its enormous surface area makes it so highly absorbent that it can strip poisons out of liquids. By then the polluted water is pure. Every few hours the big collection of liquid grey mud must be drained from the bottom of the cone.

Some towns are considering a double water supply: out of the river for gardens, filtered water for houses.

In the 1970s the alum used in these plants gave great promise in a series of experiments by Valerie May in farm dams. She was looking for some way of stopping blooms forming, instead of treating them after they had formed. In mid-winter she floated bags of ferric alum in the dams to dissolve slowly. It has the power of absorbing phosphorus, so by summer, when it had mixed thoroughly in the water, it had taken all the phosphorus lying in the sediment at the bottom of the dam and the blue-green algae that hatched had no fodder.

It is phosphorus that feeds the blue-green algae which develops in the unnatural stable conditions of our waterways, so it must be kept out of them. But even if all phosphorus influx were stopped tomorrow there is enough in the beds of all our dams, rivers and creeks to feed massive blooms for up to twenty years. There can be no quick fix for our rivers.

Where does the phosphorus come from? A great deal comes from the thousands of tonnes of phosphatic fertilisers used on agricultural crops and this indicates another great problem, that of soil erosion. Phosphorus binds immediately with clay particles in soil so it cannot wash into streams as a solution, it must travel in with crumbs of soil. Much more phosphorus comes from what are known as point sources:

sewerage, abattoirs and stormwater drains. Dog droppings in astonishing quantity fertilise stormwater drains, blood and manure from hard-packed yards are the source from abattoirs. (Fishermen have had to change their opinion about abattoir run-off. The outlet pipe into the river was always a favourite fishing spot, fish collected there for the choice food.) Sewage treatment plants are the main source and there are 171 in the Murray–Darling basin. Only three of them remove phosphates. Some towns, like Walgett, Brewarrina and Bourke, try to avoid polluting the river by pumping their sewage into lagoons where the liquid evaporates. Rainfall out there is 450 millimetres a year and the evaporation rate 2 metres, so there is no disposal problem. But only the water evaporates. The phosphorus stays behind to be dissolved during floods and much of it eventually reaches the rivers anyway. Too many towns discharge their sewage water straight into rivers. The worst towns, in order, are Toowoomba, Orange, Bathurst, Tamworth, Dubbo, Gunnedah, Moree, Inverell, Narrabri, Dalby. So the Namoi River, with three of the worst towns on it, is the most polluted river.

Fifty-five per cent of the phosphorus in the discharge from sewerage pipes now comes from laundry detergents. In the 1960s Lake Erie on the Canadian–American border became known as the dead lake. It seemed blue-green algae had killed it. At that time detergents were up to 50 per cent by weight of phosphate. Canada limited the amount of phosphorus in detergents to 2 per cent and over twenty years the lake made a great recovery. Recently the state of Ohio, still concerned about the lake, reduced the level of phosphorus in detergents to 0.5 per cent.

On 13 May 1992 Ian Causley, Minister for Natural Resources, issued a press release, headed "<u>PHOSPHORUS MUST BE REDUCED IN DETERGENTS</u>". Note the underlined capitals. This is what he proposed; it is the major statement of his press release: "At Friday's Murray–Darling Basin Ministerial Council meeting in Canberra, I shall seek support from all states in establishing the five per cent level of phosphorus in detergents nationally, by 1997." Both the 5 per cent and the delay are ridiculous. Causley has betrayed his Blue-Green Algae Task Force and ensured continuing problems. Last summer blue-green algae was reported in forty-eight waterways, in thirty of them at an extreme level of alert. For the last four years Chaffey Dam on the Peel River above Tamworth has produced extremely poisonous blooms.

It would not be unreasonable — in fact, it is necessary — to require an immediate stop to phosphates in detergents. If manufacturers protest, and they certainly would, they could be countered by advertisements listing the small companies such as Health Script Laboratories of Balcatt, Western Australia, who make superior laundry and dishwasher detergents containing no phosphorus.

Even when the sensible detergents are used, it would be preferable that no sewage water entered waterways or oceans. This is a valuable fertiliser as the Metropolitan Sewage Farm at Werribee has demonstrated for years. Some country towns, such as Grenfell in central-western New South Wales use it to spray parks and golf courses. The best use of it, a valuable use, perhaps even a long-term payable use, would be to grow forests for milling or chipping. The city of Albany in Western Australia is now planting 500 thousand fast-growing gum trees to use the sewage water which now flows into King George Sound; at Wagga in New South Wales the CSIRO is experimenting with tree plantations watered and fertilised by city sewage.

Native Australian trees can sometimes produce extraordinary growth when moved out of their natural environment. In Tasmania, Australian Pulp and Paper Mills experimented with Victorian Shining Gum, *Eucalyptus nitens*. It so enjoyed its leap across the strait that it is now known in Tasmania as Supergum. It is behaving even more vigorously than eucalypts planted out of their environment overseas. In some sites it reaches 6 metres in three years. Forests of them can be thinned at nine years and harvested for pulping at fifteen years. If the prized Blackwood, *Acacia melanoxylon*, is planted among it, the eucalypts encourage the Blackwood into quicker growth. A forest of Blackwood and Supergum together can produce three cash crops in thirty years, which makes the growing of them viable for farmers. It required a superhuman assurance to embark on 100-year projects.

Supergum is limited in range, it cannot cope with too much snow. So foresters are now experimenting with crossing it with the superb Tasmanian Blue Gum to give it a boost. They call the seedlings "speedlings".

The final report of the Task Force draws attention to the importance of the animal life in water: "Increasing the populations of the large herbivorous zooplankton, such as *Daphnia* or large cladocerans, can reduce undesirable blue-green algae." Water-fleas, shrimps, smaller crustaceans feed on blue-green algae when it is in normal numbers, so do other forms of life: fungi, actinomycetes, chitrids and amoeba. (Amoeba are minutely monstrous, described as "a microscopic one-celled animal consisting of a naked mass of protoplasm constantly changing in shape as it moves and engulfs food.")

None of this life can return until the water plants return. And perhaps the plants themselves inhibit blue-green algae by releasing a chemical as they die which is poisonous only to blue-green algae. A few years ago an English farmer noted that when rotting bales of barley straw blew into a lake blue-green algae disappeared. Interested scientists found that the straw released an unknown chemical. Other straws had it in much smaller amounts, what of Australian water plants?

A vital fact that has not been stressed is that it is our interference with waterways by dams and weirs that is the prime cause of blue-green algae; regulation is intolerable to streams which developed without any discipline whatever. Apart from its use in mining from the 1850s on, the first consideration of the extensive use of river water was for irrigation which began with a weir on the Loddon at Kerang in Victoria in the 1870s. Ideas changed with increasing river traffic. Between 1870 and 1880, 200 paddle-wheel steamers towing barges worked the Murray and its New South Wales tributaries. They brought so much profit to South Australia that the colony feared the development of irrigation; what it wanted was locks on the rivers to make travel more certain at times of low flow. It takes a long time for politicians to act. They began to build a lock at Blanchetown in South Australia in 1913 when the steamers had almost given way to trains.

The first weir and lock on any New South Wales river was begun on the Darling at Bourke in 1895. The contractors put a temporary bridge across the river to get building materials to the other side. Captain Pickhills came downriver in the loaded *Mundoo* with his barge *Duck* in tow, anxious (as captains always were) that the river would hold its level long enough for him to get to the Murray. He tied to a tree on the bank and asked the construction engineer to take enough of the bridge down to give him passage. The engineer refused, so Pickhills built up steam, hitched a rope to the bridge and pulled it down. The expensive and well-built lock was never used.

Answering a cry of "Lock the Darling ... stop the flood waters from rushing away to the sea," which city newspapers took up, sixteen more weirs were built on the Darling and Barwon, more than a hundred on other rivers in the basin. Some of them were for urgent and necessary town water supplies. In the early 1890s when Dubbo had a population of 6000, each summer saw it dangerously short of water.

All weirs must now be reassessed, the rivers cannot tolerate them. Some can be removed and town water supplied from bores, many can be lowered, all require broad fish ladders. There are very few ladders at all, those in place are insufficient. Native fish can migrate only in time of flood when fast water makes migration difficult. Fish cannot maintain a healthy population unless they can migrate, the rivers cannot stay healthy without native fish.

The Darling River is speaking for the whole of western New South Wales. The land has degraded like the river. Rabbits as a grey blanket in the 1880s, starving rabbits, sheep and cattle in a long drought in the 1890s began the devastation. They stripped the land of grass, herbage, shrubs and trees, then dug for roots and seeds. This country is resilient, say the landowners, and so it is. But elasticity is not permanent, the best springs eventually fail. Saltbush has gone from huge areas; there is little mulga where it ought to be, it is growing like a weed where there ought to be grass. Parakeelya and various pigfaces are gone from the sandy country where sheep used to be taken to over winter to give the river country a spell. Almost all the original good grasses are gone from the whole of western New South wales. Introduced Wild Sage, *Salvia verbenaca*, takes over bared ground, or the native everlasting Poached Eggs, *Muriocephalus stuartii*, spectacularly beautiful but useless as food to any animals, whether native or introduced.

Amazingly, when experimental areas of this country are fenced off, they do recover. Even grasses that no one has seen since last century come back in a year or two as though by some sort of spontaneous generation, a virgin growth without a seed, a memory taking root.

So there must be change in the management of that country. It seems that 30 per cent of western graziers will soon fail anyway. Their holdings, divided into soldier settler leases with foolishly high government-decreed stocking rates, are now too small. CSIRO scientists consider that the country will have a chance to recover only if stocking rates are reduced to one-quarter of the present rate which is now about one sheep to 5 hectares. Perhaps that land could be used as the brilliant

Kidman used it, intermittently for fattening instead of breeding, so that each station goes for months with no stock on it.

No change in stocking rate can be successful unless the extraordinary numbers of feral animals are dealt with. The Murray family at Louth yarded and sold 33 thousand wild goats in sixteen months of 1991–92. In June 1992 a count from a helicopter found 7741 goats and 974 pigs on the 48 thousand hectares of Mount Mulya at Louth. Rabbits and foxes were uncountable, there were 2600 sheep and 8962 kangaroos. The kangaroos ought to be a marketable asset. The sentimental resistance to killing them is ridiculous: their skins make the best shoe leather of all, their flesh tastes good and is low in fat, they have thrived since European settlement and are now in quite unnatural numbers. Even kangaroos can be damaging when there are too many, as Hattah–Kulkyne National Park in Victoria demonstrated. Starving kangaroos ate the bark of trees as high as they could reach, standing on tiptoe and grasping trunks with forepaws for the last mouthfuls.

The Darling River banks are barren. Certainly there are magnificent River Red Gums, Coolabahs and various melaleucas and casuarinas, but the smaller plants are gone, the soil is always exposed. Hundreds of kilometres of the river banks were stripped bare in the great sheep movements of the 1890s and they have been bare ever since, grazing pressures during droughts bared those stretches bypassed by stock routes. I know the difficulty of maintaining fences in flood country, I had twenty years' experience, but it is feasible to fence off long stretches of the Darling to allow the banks to regenerate. The highest land is often on the bank of the river — the floods spread behind it, not over it. The watercourses and channels such as those which sometimes lead the Paroo River into the Darling are probably impossible to fence. It will require wise surveyors, who know the country and who know fencing, to decide what is possible. Decisions by politicians and public servants for fencing off a standard strip would be unworkable and absurd.

A brilliant artificial solution has been determined for Carcoar Dam on the Belubula River in central New South Wales. Yvonne Goldie, a teacher at Glenroi Heights Primary School and a sailing enthusiast, too often found her yacht trying to plough through 10 centimetres of stinking slime. (There is ten times as much phosphorus in the sediment at the bottom of that dam as in the Darling River.) Yvonne formed the

Carcoar Dam Catchment Management Committee to work with Water Resources, which was considering an artificial wetland to filter the river. Blayney abattoirs, sewerage works, town run-off, rural run-off had almost killed both river and dam. Nothing thrived except blue-green algae. Conservation and Land Management joined the scheme which is now in operation.

A new weir blocks the polluted river and diverts its full flow into a 9 hectare system of earth baffles through which it zigzags slowly back into the river and into the dam, dropping its pollutants as it goes. The planting of filters and shade and food for animal life began in 1992. Yvonne Goldie organised her class as nursery workers. She set up a propagation room with heated germination benches, a local nurseryman, Graham Hawke, supplies pots and seeds. The children spend part of every day with the plants, they have come to regard the wetland as their own.

It is a fascinating social experiment as well as an environmental experiment. Glenroi Heights is classed as a disadvantaged school. Many of the children had behavioural problems, more likely to smash something than nurture it, before they began work in the nursery.

The selection of plants is excellent, not just the usual mix of trees which attract a limited associated life. Blayney High School has joined them and together they are planting ground covers and a range of trees, shrubs and herbs growing from 30 centimetres to 30 metres high. They will sow native grasses and plant sedges, reeds, rushes and lilies in the bays. Blayney abattoir, the main cause of the pollution, has agreed to install a nutrient removal system. For years they have been thought to disguise deliberate releases as broken pipes and unfortunate errors.

If this system works — and everyone associated with it is confident — more such artificial wetlands will be constructed. It would seem more sensible to restore the natural filters, beginning with the Macquarie Marshes, now reduced from 100 thousand hectares to 6000, which are suffering severely from recent banks and channels put in by cotton growers, despite a yearly environmental allocation of 50 thousand megalitres of water for the marshes. It can no longer spread as it ought to, the Macquarie River no longer flows down its original course, it flows down Monkeygar Creek, once an overflow, now a mere irrigation channel. Apart from their huge worth as a filter, the marshes are, perhaps were, one of the biggest breeding grounds for waterfowl in Australia.

Yet when Sally McInerney wrote an excellent article for the *Australian Magazine* about the sad state of the marshes, some members of the local Total Catchment Management Committee expressed anger that outsiders were getting involved in their affairs. They thought reporters should be banned from the marshes. Some have admitted they do not worry much about licences for flood control and diversion banks. They move water where they want, licensed or unlicensed. One is known to have built kilometres of unlicensed diversion banks. Landowners seem to be able to act as they please, the Department of Water Resources seems to have no power and no inclination to do anything about illegal works. Peter Millington, the Director, asked Sally McInerney: "What's the big obsession with the Macquarie Marshes? There's no doubt there are banks around that we haven't been able to tidy up, but overall it's not too bad." It is now very bad, and it is rapidly getting worse. When the marshes are finally destroyed, will he construct an artificial wetland of several hectares?

Nobody, except the operators, has yet caught up with the enormous power of modern equipment. Des Stevenson of *Cubbie* station in Queensland is a firm believer in state rights to water. An open-cut miner, he sold out and bought 88 thousand hectares straddling the Culgoa and Narran Rivers. For four years, six caterpillar D10 tractors and two D9s have been constructing irrigation channels bigger than any of the rivers, he built a massive bank 10 metres high and 20 kilometres long across the Culgoa–Balonne Minor floodplain — legal work in Queensland but not in New South Wales. When the 1990 flood was bearing down on millions of dollars worth of harvested cotton, he built a road, not a bridge, straight across the freshening Balonne Minor and Narran Rivers to a bitumen road and got the whole crop out. He has dammed the rivers to within 30 centimetres of the top of their banks and proposes to instal twelve 42 inch axial flow pumps (1.1 metres in diameter). I do not know of anybody who has yet seen pumps of that size. Everything that Stevenson has done is legal, the engineering of the scheme is brilliant.

But the Culgoa supplies one-quarter of the Darling's water. It comes down in a white stream that keeps to itself in a broad band in the Darling water for 30 to 40 kilometres before it mixes. *Cubbie* station will use most of it.

It is better that the devastation of inland New South Wales be rectified by education than by laws: compulsion yields no knowledge. But much firmer laws are necessary to control those landholders with massive money and the massive capability to destroy river systems. Farming and grazing are great occupations. No farm should be a barren display of moneymaking and engineering, even the biggest cotton farm could be a thing of beauty and an environmental asset with the right planning. That is what the Darling River is demanding.

The north coast story

The north coast of New South Wales is at once the loveliest place in Australia and the most degraded. Its population will double over the next thirty years and there is no provision to cope with 350 thousand extra people. There are the people with the knowledge and the drive to restore it and sustain it but it will be a difficult transition. They are in positions subject to political control and government, both state and local, is arguably the worst in Australia.

For almost fifty years the north coast remained secluded while squatters spread sheep and cattle over most of southern Australia. Penal colonies on the northern and southern boundaries isolated it, as did barred and hidden river mouths and mountain walls on the west. The area stretches south from the Queensland border to Port Macquarie and the Hastings River 350 kilometres away, and the east-west distance from the ocean to the Great Dividing Range averages about 120 kilometres. Over most of the western boundary this range is the great escarpment, a wondrous wall up to a thousand metres high. Not even rain clouds driven up from the sea can climb over it. They pile up against it then tumble back down towards where they came from, spilling one to two thousand millimetres of rain a year.

Clouds often drape Mount Warning who overlooked Cook when he almost ran aground at Point Danger. Lightning plays around him in such spectacular fashion that the Aborigines knew him as Wollumbin, Fighting Chief of the Mountains. It is fitting that he overlooks the area like a stone god, given that 23 million years ago he created it. There is no question of his sex: he did not give birth, he ejaculated molten granite in a supreme emission that flowed over a circle about a hundred kilometres in diameter and up to two thousand metres deep. Other outpourings followed from him and from lesser volcanoes. They destroyed most of the ancient rainforest but sufficient pockets survived to regrow the area as the lava weathered. The physical and chemical actions produced the greatest range of soils in Australia and they came to support one of the greatest diversities of plant species.

The foothills of the ranges that fed the rivers were mostly open grassland. An explorer might have stepped from dense dark forest into sunlight in a metre or two. Rivers and creeks were fed by big lakes and treeless, grassy swamps; they cleaned themselves again through vast areas of melaleuca swamp and mangroves before they entered the sea. Some creek and river banks were open, others walled in with gigantic trees that were hung with ferns, entwined with creepers, roped together with vines. This luxurious rainforest varied in width from a few metres to couple of kilometres and it edged a beautiful mosaic of open country varying in size from 20 hectares to a few hundred. The grassy flats and downs that it outlined with green were often dotted with either Smooth-barked Apples, *Angophora costata,* acclaimed as "the most picturesque forest tree in Australia", or gigantic Blackbutts, *Eucalyptus pilularis,* that grow in a fortunate band from Bundaberg to Bega. Some of the ridges in the west were grassed on top with scrubby sides, others were timbered on top with grassy sides, depending on how the Aborigines of the area had managed their land with fire.

The extent of open country on the Clarence and Richmond rivers was amazing. On the Clarence swamps became plains covered with head-high grass, one divided from the other by forested tongues of hills running down to the river. These plains averaged six trees to the hectare and there were no shrubs. Past the Broadwater on the Richmond as the river wound round to the north, a thick jungle, the Big Scrub, became visible on the east bank; the west bank opened up into big plains. For another 60 kilometres of the river's winding path, a strip of country 20 kilometres wide (120 000 hectares altogether) grew nothing but grass. "Although surrounded by trees of a hundred varieties," wrote Oliver Fry, first Commissioner of Crown Lands for the Clarence district, "still, in surveying the vastness, the eye seeks in vain for even a single shrub upon which to rest."

The Big Scrub was the chief feature of this north coast. One-tenth of 1 per cent of it now exists. It was one of the world's great features, thousands of years old and containing hundreds of thousands of the finest specimens of timbers the world has grown. It stretched from the Tuckean Swamp above the Broadwater almost to the Brunswick River where a different forest took over and ran up across the Tweed valley.

On the east it began where the swamps or sand dunes ended; the north-south Terania Creek, a tributary of Wilsons River, fixed its western boundary. It covered about 75 000 hectares and three-quarters of it, no more, was rainforest made up of at least four different sub-alliances of plants. Wet eucalyptus forest that grew in belts and patches and along ridges made up much of the remaining 19 000 hectares, there was some heath, and strange treeless areas known as Grasses occupied several hundred hectares. Nobody knows how many of them there were; some were given European names and nine of those still show on detailed maps, among them Dorroughby Grass, Molly's Grass, Middia Grass, Chilcotts Grass.

Other extraordinary growths were the narrow strips of littoral rainforest — there is still about 100 hectares left in one block at Iluka. They were dense and beautiful, an intense association of plants, each depending on other species for life. Many of the trees in the canopy and many shrubs in the second layer produced edible fruit, some of them spectacularly so. Copper laurel, *Eupomatia laurina,* produces massed cream flowers with an overpowering scent and very sweet greenish fruit with a lingering aftertaste. Pollinated by beetles which feed on an inner ring of infertile petal-like stamens, these shrubs are ancient and have no modern relations. They make one realise that the rainforests that disappeared under lava might have been perfumed gardens. Wind-sheared canopies were a special feature of these forests: natural topiary on a grand scale. There is a reasonable remnant at Broken Head. The perfectly even top of the protective canopy 10 metres above the ground follows exactly the 40° slope of the hill. Not a branch, not a twig, not a leaf juts up to break the order. Although the top of the canopy looks green and healthy, wind and salt blast off any shoot that rises above it.

Altogether the north coast rainforests have a profuse complexity of plants: more than 550 species of trees and shrubs, seventy-eight orchids terrestrial and epiphytic, ferns terrestrial and epiphytic, palms, cycads, grasses, herbs, mistletoe, fungi, lichens, mosses, and some hundred species of vines lacing them and shrouding them and sometimes hanging from top branches in marvellous curtains 6 metres long with a 10-metre drop.

The rainforests glowed with flowers. The spectacular Flame Tree, *Brachychiton acerifolius,* is now more or less domesticated. It once blazed all through the forests, springing into flower after it dropped its leaves in summer, "a mass of the brightest scarlet, enough to dazzle one's eyes". The prolific fruits were mostly blue, black or red, colours that attract birds. Many seeds cannot germinate until they have been scarified by a bird's gizzard.

The timbers of the rainforest are (were) magnificent. Eighty-three species are on display under good lighting at the Richmond River Historical Society's museum in Lismore. Prepared by Evan Williams, a former sawmiller, they are not the usual toy blocks but metre-long planks 20 to 30 centimetres wide that show the wondrous character of each piece. I am having difficulty in this article in deciding whether to write in the present or the past. Regrettably, with too many of the timbers, "were magnificent" is unquestionably the case. Some species were never plentiful, others like the fairly common Bog Onion, *Owenia cepiodora,* were simply overharvested. By 1977 there was not a single mature tree left in the wild. Except for its oniony smell, the timber has a remarkable resemblance to Red Cedar (though few of its qualities) and rascally timbergetters found that they could soak the smell out of it in running creeks then market it for many times its value with a few logs of rare cedar.

Outside the rainforests the eucalyptus hardwoods are outstanding. To my eyes accustomed to western trees, they are so big, so astonishingly tall and straight and limbless: Spotted Gum, Narrow-leaved Ironbark, Blackbutt, Sydney Blue Gum, Grey Gum *(E. propinqua)* and the popular Tallowwood, huge, longlasting and so fast growing that it is used as a windbreak for macadamia plantations.

The varied conditions supported an extraordinary range of animals. Dense wet conditions require specialist lives, but in the Big Scrub the Grasses and the strips of eucalypts gave opportunity to animals that could not cope with rainforest. So gliders, possums, Grey Kangaroos, the little Parma Wallabies and other creatures of the open grassland and eucalyptus forests lived in there as well as the Red-legged and Red-necked Pademelons that belong to rainforest. The big rainforest pigeons wear colours to match the brightest flowers and fruit but they are disappearing with the flowers and fruit. Near Kyogle there is a district known as Green Pigeon, named for Greenie, the Wompoo Fruit-Dove. Its species name means "magnificent", but the district name is now mere remembrance, there are no more there. Brownie, the

Brown Pigeon, works all levels of rainforest and eucalyptus forest and uses its long, strong tail as a lever and an arm to sweep leaves out of the way when it is feeding. The lovely Emerald Dove, mauve and iridescent green, feeds on the ground.

Two extraordinarily rowdy birds, the Noisy Pitta and the Eastern Whipbird, spend most of their lives on the ground. Pittas have such bright, sharply distinct colours — green, buff, blue, red, black, chestnut — that one would expect them to be easy to see against any background, yet they can lose themselves among green leaves and forest litter even when their shouts of "walk to work" tell where they are. They are noisy, too, in raking the ground for worms, beetles, slaters. Especial delicacies are the several species of native snails which they crack open on a favoured stump or stone. Repeated use wears these anvils smooth, a deep midden of broken shells surrounding them.

The big, bright Richmond Birdwing Butterfly with a wing span of 18 centimetres almost disappeared with the Big Scrub but it was fortunate in being dependent on one genus of plants only, two species of a vine, *Aristolochia*. A cultivated exotic of the same genus is the well-known Dutchman's Pipe. So it does not need forest to live successfully — vines planted for it in gardens will ensure its survival.

The value of even small areas of rainforest is demonstrated in a patch on Upper Coopers Creek Road, north of Lismore and east of the Whian Whian State Forest. David Milledge, a well-known zoologist, lives beside a lovely, steep 40-hectare rainforested valley. In it he has seen Albert's Lyrebird, a marvellous mimic, as well as Marbled Frogmouths and Sooty Owls, two very rare nightbirds. Powerful Owls sometimes pay a visit. There are also two rare frogs in that pocket, Loveridge's and the Pouched Frog. Living in wet litter, not water, and only 30 millimetres long, they had original overlapping ranges of no more than 170 kilometres by 140 kilometres, mere blobs on the map of Australia, so they have been much affected by clearing. Males of the Pouched Frog have pockets along their flanks in which they carry the fertilised eggs.

The streams, estuaries and swamps were rich beyond present comprehension. The lower Richmond and especially North Creek held Sydney Rock Oysters in enormous numbers. The Clarence estuary was reported as abounding in fish of enormous size, many of them 1.8 metres long. They must have been Greasy Cod, also known as Estuary

Cod. When tides were not interrupted by floodgates, flathead moved astonishing distances upstream. Even this century they have been taken where Terania Creek flows into Leycester Creek 5 kilometres west of Lismore.

There were Murray Cod in the Richmond and Clarence tributaries but they are now confined to the upper tributaries where they are protected. There are still Australian Bass in upper streams, too, but these excellent eating and fine angling fish cannot tolerate any interference with rivers. The middle and lower stretches of rivers became uninhabitable when protective logs were cleared for the passage of riverboats and log rafts: flood gates and weirs interfere with spawning runs and prevent the entry of young fish that hatch at sea then swim back into the nearest river.

One little fish of fast mountain streams and waterfalls is not much troubled by weirs. Using spiny, well-developed pectoral and pelvic fins, Cox's Gudgeon can climb wet rocks. It has been seen right out of water climbing up the damp, vertical concrete face of a weir.

The rich supply of food sustained several thousand vigorous and intelligent Aboriginal people. Good-looking and of imposing height — many topped 1.8 metres — those Europeans who first saw them admired their appearance, their craftsmanship and their intelligence. To recognise Aborigines as people was rare enough; this was recognition as a people to be considered.

There were probably thirteen main tribes or language groups but each of them divided into clans or sub-tribes of eighty to a hundred people speaking dialects based on the main language but often differing so much from it that one clan could not speak to another. They built more permanent homes than was usual, living for months at a time in semi-permanent neat villages of thatched circular huts up to 15 metres in diameter and 2 metres high.

They built involved stone fish traps in rivers and on rocky headlands in the sea and used brush fences as temporary traps in creeks. Their twine making, net making and basket weaving were beautifully done. The women fashioned deep, boat-shaped water carriers out of the leaves of the Bangalow Palm by bending the main rib of the long, pinnate leaf to the shape that they wanted, tying the ends together with a cord handle, then lacing the pinnules in such a tight weave that no water seeped through them.

The men made the net tow-rows that they used for catching fish. Knotted to the same ancient pattern as modern European nets out of cords rolled from the fibres of kurrajongs and stinging trees, these were in the shape of an oval about 2.5 metres across, kept in shape by springy bamboo spreaders to which they attached a handle. Lines of men took one in each hand, surrounded shoals of fish in waist deep water, drove them into bank or beach and scooped them up in the nets.

Although still suffering the devastation of displacement and the effects of the change from a very healthy Aboriginal diet to the worst possible European, the Banjalang people are one of the leading black races. In the mid-1970s they instigated special classes to teach their languages to their children, and organised classes for adults in useful trades. On 23 May 1991 they published the first edition of the *Koori Mail,* "The Fortnightly National Aboriginal and Torres Strait Islander Newspaper". The Aboriginal flag in black, red and yellow flies on its masthead.

The destruction of the Aboriginal people was a tragedy for the land as well as for the people. As they numbered no more than 9 thousand, there was plenty of room for black and white. Twenty per cent of forest, grass and swamp left in the hands of Aborigines would have maintained both land and people. As early as 1842 a leading Aborigine implored Edward Ogilvie, one of the first squatters, to leave his tribe alone in the mountains where they had retreated. Ogilvie lost something of the passion of the plea by recording it in archaic language. "Begone, begone and take away your horses, why do you come hither among the mountains to disturb us? Return to your houses in the valley, you have the river and the open country, you ought to be content ... go back — keep the plains and leave us the hills."

What happened was inevitable. In all settlement in Australia wild cattle and wild men moved well beyond the restrictions of law and domestication. Civilisation was a staid, distant structure. In his report from the Macleay River in December 1844 Robert Massie, Commissioner for Crown Lands, considered the difficulty Aborigines would have in assessing white behaviour: "Their first knowledge of civilized life is gained from the intercourse with such lawless and unprincipled men as are generally ... the forerunners of civilization in this colony."

The official attitude of concern for the Aborigines was to Christianise them and teach them agriculture. They had no need of either pursuit.

Cedar getters made the first real intrusion into their territory. Red Cedar, *Toona australis,* was named for the resemblance of its timber to that of the *Cedrus* genus, Indian cedar or the cedars of Lebanon. Written of in biblical times, regarded as ancient, the cedars of Lebanon were contemporary with Australia's Red Cedar, thousands of years old when harvested. The eucalypts about Sydney were a great disappointment to the new settlement in need of building timber. Knotted, misshapen, twisted, ruined by fire, they were of little use. Then Red Cedar was discovered on the Hawkesbury river. It looked magnificent and was easy to work. The government claimed title, sent gangs of convicts out to cut it. Samples went to England, shiploads went to England and to China, India, New Zealand, South Africa, L'île de France (Mauritius). The timber gangs cut out the Hawkesbury and moved on to the Hunter River. Whole houses were built of Red Cedar, as well as public buildings, fine furniture, packing cases, paling fences. There was so much of it that it could be used as deal. Cedar cutters moved down the coast, up the coast. Members of the convict gangs worked out their sentences then cut for themselves, free settlers joined them, there was money to be made. And to avoid licence fees, the cutters kept as far ahead of the law as was practicable.

Red Cedar loses its leaves in winter, flares with copper-coloured new growth in the spring. It grows with other soft-leaved trees in rainforest brushes, usually as individuals. A good stand averaged one to the hectare over numerous hectares, then they spread out one to 5 hectares, one to 10. There might have been odd half-hectares with ten trees on them. Mature trees reached an astonishing size — 50 metres high and 3 metres in diameter at breast height was not uncommon. The cedar getters developed a special technique. Crosscut saw, axe, standing boards were their total equipment. Cedars are buttressed, the bases are often hollow, so they felled them at heights up to 6 metres off the ground. They would cut a slot, they called it a scaffold hole, about shoulder height, drive in a board, stand on it, cut the next slot and drive in another board, perhaps another and another. But since boards were heavy and awkward to carry through brush, many used one board only.

After they had cut the second slot they would cut a toe hold to stand in while they moved the board up.

They worked the rivers one by one, the Hastings, Manning, Macleay, the Clarence by 1835. Only the "handy" timber could be taken at first, that near enough to rivers and creeks to be levered into the water and floated down to waiting ships. Aborigines were good to the first comers, showing them where their many tracks led, offering food, women. The cedar getters were not interfering with their way of life, nor were they destroying the land; they were taking one species of mature tree with minimum disturbance to the forests.

It was squatters who upset the land and the Aborigines. By grant and by purchase settlers had begun moving on to comparatively small blocks on the Hastings soon after it opened as a penal settlement. By 1825 farming had extended 30 kilometres north-west to the rich and easily worked Rollands Plains. But in 1826 sheep of the huge Australian Agricultural Company took over the Dungog, Stroud, Gloucester country and the Manning River. Then in 1832 H.C. Semphill squatted at Wolka (Walcha) on a run that tipped over the eastern watershed to the head of the Apsley.

In 1836 Major Archibald Clunes Innes moved a few thousand sheep to a squatting run on the south bank of the Macleay. Other squatters soon joined him, bringing their sheep through the country that the AA Company had opened up. By 1837 about thirty squatters had taken country north of Wolka and their shepherds were looking down on the Macleay river, among them those working for Major Innes who had taken another run at Kentucky. By 1839 there were 200 "graziers, mechanics, farm servants and mariners on the Clarence". Richard Craig, a young convict runaway and a truly remarkable man who had lived with Clarence River Aborigines and learnt their language and their country, engaged to bring squatters' sheep and cattle off the New England down to the lower Clarence. Early in 1840 he took the first mobs in, blazing a track that became known as Craig's Line, burning the high grass in front of them so that they could get the sheep through. Part of that track is now overgrown and no longer negotiable by sheep or the drays that they sometimes manhandled from tree to tree with block and tackle.

Within several months the Ogilvie brothers had marked an easier more northerly route to get through to their Clarence River run and nearly starved to death in doing so. Thomas Hewitt, superintendent of a run near present Glen Innes, travelled north up the Dividing Range to the big Tenterfield run, then blazed a track down to the Clarence that the Bruxner Highway follows for long stretches.

During 1841–42 convict gangs put a road through from Port Macquarie to Walcha. The whole of the north coast was open by land and by sea. Astonishing numbers of sheep were driven over the several routes. The Ogilvies on Yulgilbar (on the Clarence near present Baryulgil, north-west of Lismore) were soon shearing 28 000. But the climate was too wet, the grasses too long. Whole flocks died of diseases now eradicated or treatable: "fluke, bottle [worms], foot-rot and catarrh". Some squatters moved to Queensland, and those that stayed gradually switched to cattle.

In the spring of 1841 (not 1842 as is often stated) an experienced, energetic and restless cedar cutter named Steve King led a party from the Clarence to the Richmond. Aborigines had told him that there was good cedar there. Steve King's Plain south of Lismore is still marked on maps. It carried the clumps of cedar that cutters dreamed of, a fabulous stand where a man working on one tree might have three or four more in his sight. There were soon hundreds of men and a number of women on the Richmond. The Nambucca opened, the Bellinger, then the Big Scrub.

Distressed Aborigines began droving away flocks of sheep from those who still persisted, they speared cattle and shepherds. Avenging parties rode out and shot men, women and children, up to a hundred at a time. Some sawyers and squatters got on well with Aborigines who from the earliest times worked as shepherds and stockmen. They built roads and fences, they worked in sawpits, they carried stores and mail. Numbers of half-caste children were born — over thirty years they totalled a few hundred — and all were allowed to live, an unusual thing among other Aborigines. By the mid-1840s squatters and selectors had taken up the whole of the open country, the good, the bad, the impossible such as Matildadale, a stretch of coastal heath between the Sandon River and Corkscrew Creek which runs into the sea near present Wooli. The area is now the beautiful Yuraygir National Park but as late as the 1890s a landowner was trying to run cattle on it. "It is a very poor country," reported the stock inspector, "subject I believe to coast disease and certainly not of a fattening character." The insidious coast disease was caused by deficiencies of copper and cobalt.

Every run encompassed forest with grassland. That was not a policy based on wisdom — the nature of the country allowed for no other division. Map after map shows big areas marked "high steep ridges", "dense pine and cedar scrub", "steep grassy ridges", "thickly timbered, scrubby and broken", "scrubby mountainous country". It is fascinating to compare the excited description of landholders of the runs that they took up in the 1840s with the calculated, uninvolved assessment of 1880s inspectors, especially Charles Ralph Blaxland, stationed at Casino, who had no enthusiasm for anything that he saw. In her memoirs Mary Bundock told of childhood at Wyangarie and her father's first sight of the run in 1842. "He went up the river in search of the dry open country described by Edward Ogilvie, and one afternoon rode over the low gap which leads to the lovely Wyangarie plain. He thought it the most beautiful spot he had ever seen, a smooth open plain with one clump of heavy timber and two or three small lagoons sparkling in the sun ... the whole country was covered by a thick crop of kangaroo grass then in seed, and looking like a crop of oats."

And after thousands of big Shorthorn cattle had carried B for Wellington Bundock branded on the nearside rump for more than forty years, Blaxland wrote, "There are about 31 680 acres of jungle consisting of tall trees imperviously matted with vines, cane, creepers etc. and containing nothing that stock can subsist upon". He did not describe the open country where the cattle fed, he simply stated its area which was much the same as the jungle.

For holding after holding in the three pastoral divisions of New South Wales, eastern, central and western, land inspectors recommended ringbarking as an improvement. This added assessable value to a property because it did increase the growth of grass for a time. There was a strange contradiction in what was happening to the land. While trees were being cleared on some runs, on others young trees were starting up, open forest was becoming thick scrub. It was caused by the disruption to Aboriginal burning, to the disturbance of the soil by cloven hooves — it was a protective reaction by troubled soil.

And the original grasses, suffering under cloven hooves, unable to thrust new roots into hardening ground, were giving way to the vigorous and inferior *Aristida* and *Stipa* species as they did everywhere else. The original grasses were dominantly Kangaroo Grass, *Themeda australis,* which was the principal grass from southern Tasmania to the northernmost tip of Western Australia, and Blady Grass, *Imperata cylindrica,* which is the Kunai of Papua New Guinea. Cattle and sheep (until worms and footrot and the wet overcame them) thrived on grasses that grew in special niches of their own or mixed with the dominant species. They were common enough to be important, now they are overcome by introduced grasses. There were some forty or fifty of them, a significant number, that built the early north coast cattle. It could take days now to find a single plant of some species. I doubt that any are extinct. Grasses have a supernatural ability to survive.

Cattle breeding became a stable pursuit. Distances that stock had to be moved were of little concern, it was a normal part of running a property. These days a farmer makes a lot of calculations before he trucks cattle 500 kilometres. To the Barker brothers on the Ettrick run in 1845 that distance was an advantage when they bought a mob of cattle on the Hunter River "about three hundred miles down the country, which is a very convenient distance, as they are apt to stray if they come from much nearer". Until cattle got used to new country stockmen yarded them at night and tailed them on feed.

Along with the cattle industry cedar getting became well organised. It was no longer a disreputable occupation, a hideaway for law dodgers. Educated men came in with their families. The Richmond River Historical Society has a fascinating diary of cedar cutting kept by Richard Glascott between 1864 and 1877. It is the only record that I found of the hard, skilled work in getting cedar to market. In May 1865 he listed the days he spent on one tree, beginning on 13 May, "Clearing away rubbish from about a tree came into camp afternoon". Since almost every rainforest tree was festooned with vines strong enough either to prevent them falling or to pull them in the wrong direction, axe, machete and scrub hook had to be used to clear them away. After that job he spent three and a half days felling the tree and cross cutting it into lengths, and another five and a half days rolling them to the creek bank and squaring them. The tedious squaring was a necessity as ships could not handle round logs. They could roll in the hold and capsize a ship and were too dangerous to load and unload.

Whenever there was enough water to float them, the logs on the bank were "canted in" to keep them moist, even if there was not enough water to "run them". Sometimes a sudden rise fell too quickly, as on 29 May 1875. "Out today running cedar in Skinners Creek there was about 25 feet [7.5 metres] of water down the creek today it had fell 15 feet when we got there got humbuged by the timber blocking itself so did not get far."

There was a permanent massive rope stop at Lismore with floating logs tied to it to catch all loose logs. For years Lismore was known as "The Rope". Sometimes old anchor chains were used as stops. The Bundock stop in the Richmond River several kilometres west of Casino was hand forged by a blacksmith out of very heavy iron. The oval links were 35 centimetres long, rounded on the outside but beaten out and flattened on the inside to prevent the chain twisting.

When a cutter had about a hundred of his branded logs at a stop, he went out with a mate or two and formed them into a raft. Into the top ends of each log they drove eyed spikes (dogs), taking care to keep them the same distance apart. Then they moved the logs together in a long line, running chains through the eyes as each log floated into place. The chains were many metres long, it was awkward, dangerous work. When the chains ran through the last pair of eyes, they were tightened (toggled) until the logs drew together immovably. Then they erected a ridge pole in the centre of the raft and threw a tarpaulin over it to keep bunks dry, and rigged up a fireplace by laying down bark and covering it with 60 centimetres of soil. Some professional rafters built more comfortable quarters on each raft and took wives and children, even milking cows, with them. It was about a nine-day trip from Lismore down the Richmond to Ballina where the logs were shipped. They had to work the tides, tying up on the flow, travelling with the ebb. The fish they caught supplemented damper and salt beef.

By the mid-1870s the approachable big trees 2 to 2.8 metres in diameter that took days to handle were all gone. Glascott was felling two or three a day of what he called little trees "about 2200 each". The crosscutting and squaring of three of that size took him two days only. But they would have been a metre in diameter at chest height, and since cedars do not taper quickly they would have yielded three logs 3.6 metres long and roughly 67 centimetres square worth $140 000 today.

Increasingly cedar getters became farmers. More and more settlers came in. An astonishing number of sailing ships and steamers had come on the north coast run to cart cedar, so farmers could get supplies in and produce out. There were many wrecks on the dangerous bars to all the rivers. A safe entrance to the Macleay was always doubtful as the bar kept shifting. Only small ships could get into the Brunswick and sometimes captains waited a week off the Richmond bar for seas to calm or tides to rise. When the wind was against them, sailing ships hauled themselves in with anchor and rope. Some years even that was impossible. Up to forty ships have waited weeks to get out of the Richmond.

Even in the early 1850s many ships carried produce as well as timber, potatoes by the tonne, wool, hides, maize and casks of salt pork. For years maize was the chief support of timber in north coast production. It did not suffer from the disease rust which could turn wheat crops into red mush. In 1826 settlers with convict labour had grown 200 hectares at Port Macquarie yielding 7 tonnes to the hectare. Pineapples and bananas showed promise; cabbages and turnips grown for pigs yielded almost twice what these crops produced in England. Moreover, without a real winter, two crops could be taken off in a year, unthought of production. By 1869, 3000 tonnes of maize a year was coming off the Richmond; in 1888 settlers on the Macleay grew 850 000 tonnes. There was general overproduction. Sometimes pigs were turned into standing crops.

But by then sugarcane had become the next hope. Many farmers had had experience with growing it for their pigs and house-cows. So it was well known that it did not like the flooding and waterlogging of low country. Those trying commercial crops planted it on slopes and ridges, putting the first canes in by what they called the "dab method", poking the canes into holes made with a dibble stick.

Turning the cane into sugar was even more difficult than finding out how and where to grow it. Some growers chopped the stalks up with tomahawks and boiled the pieces. They put in the wrong amount of lime to neutralise acids and settle the solids and the sugar was inedible, worse than the dark, sticky stuff from the West Indies that they knew as "Kanaka juice". Others built horse-driven crushing mills of upright ironbark rollers sheeted with iron. They produced good juice but could not perfect the crystallising processes. In 1869 the Colonial Sugar Refining Company (now CSR Limited) opened its first mill on the Clarence. By 1875 the Richmond valley alone had 2600 hectares under cane, some of it in the Big Scrub. A few farmers had begun to clear the outskirts.

More mills, big and small, successful and unsuccessful, went in. Molasses and treacle became a common feed for pigs and dairy cows. Hundreds of kilometres of Brush Box tracks were laid to carry cane bins to mills, to wharves. The timber is elastic and iron wheels did not cut into it. Droghers, built where they were to work, began to ply all the rivers to cart timber and cane. Sturdy, steam-powered, flat-bottomed, drawing less than a metre of water when fully loaded, they could ignore the tides and run the rivers twenty-four hours a day, sometimes pushing and pulling half a dozen barges.

Farm workers and settlers came in from all over the world. Every district had pockets of different languages: Gaelic, Italian, German, Hindi, Chinese, Melanesian. Inevitably more and more settlers began taking a hard look at the Big Scrub. Was it possible to clear such a tangle? The deep red soil promised wondrous crops. Settlers went in with axe, saw and scrub-hook. They slashed the vines off the trees, sawed down the trees, left the tangle to dry for several months, then fired it on a hot day.

The clearing of the Big Scrub was a far easier job than settlers had faced in eucalyptus forest where trees had had thousands of years practice at recovering from fire or other damage. Destruction of the Scrub began seriously in the 1880s and by 1900 it was gone — timber that would now be worth hundreds of millions of dollars was destroyed.

The work led at first to incredulous disappointment. Farmers broadcast the seeds of ryegrass, cocksfoot, white clover expecting them to outyield the south coast pastures where they had been grown. What the accumulated vigour of thousands of years pushed up were Wild Tobacco, Groundsel, Spear Thistle and poisonous Inkweed. The Big Scrub soil was in a bewildered state. For thousands, perhaps millions of years, it had not seen the sun, it had not been hammered by direct rain. The loose, 60-centimetre deep mulch over the red soil yielded up all its moisture in a few days following rain, it hardened a little with each fall. Only the sturdiest seeds had a chance of growing.

The expected massive yields of sugarcane did not eventuate. The clearing of the trees not only let in the sun, it let in the occasional frost.

Then in 1892 Edwin Seccombe on a Big Scrub block east of Lismore noticed a few plants of a new vigorous grass among Japanese clover that he had sown. He harvested seed and took it to the Lismore Council who sowed it in a paddock next to their chambers — it is now the croquet lawn. The grass, *Paspalum dilatatum,* grew prodigiously; here was the pasture for the bared Big Scrub. It produced a new industry, dairying, that doubled the population in three years. By 1900 more than 2500 tonnes of butter were coming off the north coast, by 1907 14 500 tonnes. But the paspalum that almost hid the backs of cattle eventually became sod bound and small farmers did not have the equipment to rip it up. The massive growth also exhausted the soil and the inferior Carpet Grass, *Axonopus affinis,* took over. Dairying became constant, low-paid, monotonous work. An 8-year-old child might milk ten cows before walking 5 kilometres or more to school. In 1923 a farmer at Alstonville with a wife and six children told a select committee inquiry into unpaid family labour: "I have been dairying for nineteen years ... I have practically seen nothing of life except drudgery on the farm."

Then in 1931 the Milk Board established delivery zones and excluded north coast farmers from the most profitable market. Many dairy farmers changed to uneconomic beef production on their small farms. On worn-out pastures red cattle grew the yellowish coats showing copper deficiency. Most of the cattle were undersized and light-boned due to lack of phosphorus. But, if bought as weaners and taken to the rich pastures of Victoria or the New South Wales wheatbelt, they grew astonishingly. Big sales of north coast weaners became yearly events.

As damaging as the almost total clearance of the forests was the simultaneous draining of the marshes. They occupied considerable areas, 24 000 hectares on the lower Macleay alone, even more on the coastal strip between the mouths of the Clarence and Richmond where an observer looking down from a hilltop in 1848 reported a swamp 50 to 60 kilometres long. Apart from taking up land that farmers and graziers thought they could put to better use, swamps were regarded as unhealthy. And so they were, though the malaria and dengue that they caused were attributed to "miasmic vapours" instead of mosquitoes. In the 1880s farmers on the lower Macleay with individual areas of swamp varying from 8 to 400 hectares formed drainage unions to carry the water to the river. A flood in 1893, increased in speed and volume by the rush of water out of the drained swamps, cut a new exit into the sea near South West Rocks, 9 kilometres south of the old 1.6-kilometre-wide entrance beside Grassy Head. The river lost its meander to the sea where it dropped its mud in marsh and mangrove swamp and rushed unfiltered into the sea. The government cut a channel 600 metres long and 75 metres wide to speed it on its way. The old river mouth silted.

Work began on the Richmond mouth in the 1890s. To make its entrance safe, the government built north and south training walls to direct the river out to sea south of the bar. The river was not as easily trained as the engineers expected and it was 1911 before the work was completed and the flooded muddy river could make a quicker exit. Twenty-five years later the Tuckombil Canal was built to give more control of the flooding of the lower Richmond flats. It carried Richmond floods into the Evans River which was mostly a tidal estuary, unequipped like the Richmond with swamps and mangroves to slow down water and filter it. It poured the muddy water straight out to sea.

A notable feature of all the early streams was the purity of the water. The water that the immense complex of swamps filtered was barely discoloured. It had all been prefiltered in forests and spongy grassland soil. There was little erosion in the heaviest rain. After disturbance by cloven hooves, great slabs of the steep grassy mountainsides began sliding down on to newly cut roads. There are still frequent landslides. Muddy water ran from all the cleared forests that had never lost topsoil before. It still does. Even if the thousands of hectares of drained swamps were now functioning they would have an abnormal amount of mud to sift out.

The drainage of the swamps has been a disaster. It has destroyed — it is still destroying — rivers, lakes and sea. It did not provide extra cattle feed as expected, it reduced it. But cane farmers found the drained swamps ideal. They grow magnificent crops on them, 250 tonnes to the hectare and more of high yielding cane. But one industry cannot be allowed to exist at the expense of everything else, of fishermen, oyster growers, tourists, of ordinary life on the rivers, especially since sugar is in worldwide overproduction.

The drainage is government enforced by Drainage Acts and the rate of drainage is calculated. A fairly general working figure is that no more than 60 centimetres of water can remain on an area for seven days after a flood. The system of works to ensure this on all the rivers is massive. There are 40 kilometres of drains on the Macleay, forty-seven separate structures equipped with 210 flood gates to allow fresh water out and prevent sea water coming in. On the Tweed River, from Murwillumbah to the sea, a distance of only 20 kilometres, there are thirty-seven banks of flood gates marked on 1:25 000 maps. Every map of every river has straight blue lines drawn on it terminating in the words "flood gate". The big Belmore swamp on the Macleay once covered 1300 hectares and filled for six months of the year; the remnant now fills for six weeks.

Not even the remnants of the Tuckean and Newrybar Swamps in the Richmond valley can fill normally. Parts of the creeks that used to drain them still follow their old course but they have been deepened and the dredged soil has been used to build the banks higher. Only big floods can now get out to fill the swamps. They are too dry for too long and their nature is changing.

They were wondrously special places. Two species of trees marked them, sometimes growing together but usually as pure stands: *Casuarina obesa*, Swamp Oak, and *Melaleuca quinquenervia,* Broadleaf Paperbark, which broke into magnificent cream blooms each autumn. They often stood 5 thousand to the hectare and when the swamps were drained and the big trees cut down, small boys delighted in walking from one side of the old swamp to the other without stepping off the trunk of a tree. Both trees are extraordinarily adaptable even by Australian standards. They can stand flooding with fresh water for months, with seawater for weeks, they can thrive in a range of conditions from extremely acid to neutral, they can withstand months of drought.

The swamp water, some coastal lake water, is tea-coloured, stained by the tannins in the peat that forms the bed. Some of the beds are 30 metres deep, but the usual depth is 3 to 6 metres. A peat swamp is a biological library. In saturated airless conditions, plant and animal tissues synthesise instead of decomposing. Moreover, in the still water, they do not mix together. In those beds unaffected by changes in sea level, pollens of 150 000 years ago are in perfect condition across the layer of their era.

The marshes provided excellent summer cattle grazing, they were heavy carrying and fattened big numbers of stock. But they were difficult to manage. Cattle could get cut off in sudden floods, fences washed down, in winter the ground was a bog. With the draining of the swamps, farmers expected easy management and months more of good feed. What happened was that the good grasses died, Couch, *Cynadon dactylon*, Pinrush, *Scirpus nodogus*, and Smartweed, or Waterpepper, *Polygonum hydropiper*, took over, making a poor, unpalatable pasture. Smartweed often dominated, massing 1 metre high over hundreds of hectares. Aborigines used it to stun fish. They bruised the leaves, threw them into a pool and the fish came to the surface. It has the power of deoxidising water but it probably also contains a poison, it hurts to brush the inner lip with a bruised leaf.

There have been massive fish kills from Smartweed-impregnated water pouring off swamps. The lower Macleay regularly suffers from release of water from Belmore Swamp. In one monitored release that killed thousands of fish, the river had still not recovered five months later. Only four species of the seventeen that were normally there could be found.

Even more dramatic is the runoff of acid water carrying aluminium. Twenty kilometres of the Tweed dies at a time, the once beautiful Cudgen Lake south of the Tweed is sad, sterile water. An oyster farmer, Frank Knudson, who once fattened his Brunswick River oysters in the Richmond, has now had to leave them to mature more slowly in the Brunswick. Acid runoff from North Creek killed $80 000 worth of his oysters during 1992–93. One hundred and fifty years ago, hundreds of Aborigines gathered at North Creek for their yearly feast of oysters fat with spawn.

Two species of seagrass grew thickly in estuary water 3 metres deep or less. Seagrass is the pasture of the sea: it feeds, it protects, it supports a vigorous associated life that in turn supplies much food. Since it is a true grass it depends on photosynthesis for life, so it must have clear water as acid water kills it suddenly. When seagrass dies there is a drought in the sea. The catch of prawns, Black Bream, Luderick, Sand Whiting, Dusky Flathead and Sea Mullet has declined by half since 1955.

These are obvious effects. The sub-lethal effects that might eventually be even more damaging have barely begun to be studied.

In the early 1970s CSIRO published papers warning of the danger of acid runoff. It took until the 1990s before there was full discussion of the weird phenomenon of acid sulphate soils, though Le Franc van Berkhey, a Dutch scientist, had described the soils in 1771. He called them *katteklei*, cat clay, "demonic soil", since cats have always been associated with witchcraft in Europe. The soils, built up from plant detritus, are high in iron compounds, high in sulphates and aluminium. When covered in water there is intense chemical and bacterial action directed at maintaining the usual state of the water, sometimes acid, more usually a neutral 6.5 pH.

When drained the ground cracks, shrinks about roots, oxygen pours in and there is immediate chemical and bacterial action. Many of these peat beds contain 10 per cent of pyrites (iron disulphide) and up to 40 000 parts per million of sulphide sulphur. Aluminium silicate is always present in clay, that is what makes it feel sticky. As the pyrites oxidise, sulphuric acid forms in astonishing quantity. If there are shells in the peat, their lime neutralises the acid, usually the acid simply stays in the soil and the first flood dilutes it. Sometimes there are such extraordinary concentrations that no amount of floodwater dilutes it very much — drains run with water at a pH level of 2.5, it is almost battery acid.

It builds up behind flood gates, bursts them open and pours into rivers so fast that it does not mix but goes down as a band of acid. Oysters gulp it, seal themselves tightly as protection and the trapped acid destroys them. The condition was labelled QX disease in 1972, Q for Queensland and X for unknown. It is only recently that the cause has been determined.

Fish with acid scalds are attacked by a fungi and develop what is known as Red Spot disease, a nasty, ulcerous mass. Repeated bands of acid can destroy almost every living thing in a river, as has happened several times in the lower Tweed. The word "almost" has to be used, because what is left alive are mosquito larvae, and since there are no little fish to eat them, they hatch in unnatural swarms.

There was a particularly stupid extension of cane growing in the mid-1970s when the price of sugar rose extraordinarily. The big Newrybar Swamp west of Lennox Head that had been partially drained in 1908, then used for cattle grazing, was cleared and drained completely for cane. Most of the soil was acid sulphate, as the rotten egg gas smell of hydrogen sulphide suggested as soon as a drain was dug. "The whole bloody lot of it smelt," said Bob Smallcombe, who straightened the little creeks and put the deep, wide drains in. It is the water off the Newrybar canefields that is sterilising North Creek and killing Frank Knudson's oysters near the mouth.

The farmers back up the drains with levee banks to protect each field from floodwater. A winter flood through cane reduces the sugar level, the water in a summer flood moving slowly through a field can get so hot that it cooks the cane. So if a flood overtops a bank that is not high enough and fills his field, a farmer is likely to go out with tractor and blade and open the bottom bank to let the water out on to his neighbour's. That neighbour comes out with his tractor and blade and restores the bank, the first farmer opens it up again, so the second farmer comes out with tractor, blade and shotgun. Board meetings are seldom amicable, farmers scream at one another. They keep solicitors busy.

Instead of doing something about breeding grounds, various organisations, especially recreational fishermen, are trying to stop commercial fishing of some species, especially tailor and flathead. That would solve nothing more than who catches the fish that are left. More than sixty per cent of commercial fish depend on estuary mangroves for some part of their life cycle. Mangrove swamps have to be restored, not destroyed for canal settlements or prawn farms or simply to open up a view. There can be no extension of cane farming if the rivers and the sea are to survive. At least 20 per cent of drained swamps will have to be restored to have any significant effect, which means that many cane farms have to go back under water. The farmers will have to be compensated fairly, they cannot be blamed for the damage that they have done, which was approved, usually encouraged, by government.

What has to be realised is that tea-tree and mangrove swamps are immensely valuable. No form of agriculture has any hope of improving on the natural production of fish, crabs, prawns and shrimps. A 1990 estimate by New South Wales Fisheries puts the return from mangrove swamps at $8380 a hectare a year. Harvesting costs only come out of that.

Craig Copeland, a habitat biologist with New South Wales Fisheries at Wollongbar east of Lismore, has already begun amazing work in reclaiming swamps. Other scientists are about to begin similar work. It is vital that the old definition of reclamation be changed. To turn a swamp into a cane farm or a golf course or a housing estate is not to reclaim it but to destroy it.

Already Copeland's work has attracted the attention of scientists overseas. Professor Jay Zedler, Director of the Pacific Estuarine Research Laboratory at San Diego State University, wrote of "the benefit to the world's oceans ... the magnitude of the proposed restoration ... is highly significant". Zedler is working on several hectares, Copeland on several thousand. Two of his main projects are in the Hunter Valley but on the north coast he has begun work on the Yarrahapinni wetland which was part of the system of the old Macleay mouth. He will restore more than 100 hectares of mangroves, 400 hectares of saltmarsh. The refilling of 400 hectares of the once great Tuckean Swamp is about to begin. That is about 10 per cent of the original area. If 800 hectares of swamp and 20 per cent of the forest that once surrounded it could be restored, farmers would be able to grow crops and cattle without devastating the soil that produces them. The land would be secure again.

Unfortunately its security depends too much on shire councillors, who have little knowledge, much authority and short-sighted and expedient solutions to all problems. A measure of the concern of thousands of people is the number of organisations that form to fight various projects. There have been at least thirty over the last few years, such groups, big and small, as Clean Seas Coalition, BEACON (Byron Environment and Conservation Organisation), Australian Association for Sustainable Communities, the Big Scrub Environment Centre, NEFA (North-east Forest Alliance) and environment centres in many towns which are staffed by dedicated volunteers.

But increasingly councils are defending their actions in the Land and Environment Court instead of engaging in sensible discussion, and what is worse they are meeting justified criticism with actions for defamation in the civil court. This seems to be overkill and arguably a scandalous misuse of ratepayers' funds. There was a case involving Ballina Shire Council before the Land and Environment Court in 1989. Australmin Holdings Limited applied for permission to mine sand on the comparatively new cane land on the former Newrybar Swamp. Australia is the world's major producer of mineral sands. The products ilmenite, zircon, rutile, monazite, leucoxene, synrutile and several rare earths are precious, they enhance life. As elements or as compounds they are involved in flux coating for welding rods, in replacing dangerous lead as a pigment in paint, in providing the colour in television screens and the enormously strong lightweight metals used in aircraft manufacture, and in the production of painkillers, radiation filters and superconductors that have no resistance to electrical current. They will eventually make the ceramic engines which will produce no exhaust gases.

But the company submitted an environmental impact statement to the Ballina Shire Council which concerned the majority of people who read it. As well as private individuals the Soil Conservation Service, the Department of Planning, the Department of Agriculture and National Parks and Wildlife all questioned various aspects of the project. Half the councillors took no notice of their letters, one councillor admitted that he had not even read the Environmental Impact Statement. On 19 January 1989 the council approved the scheme. But the vote was equally divided and Keith Johnson, the president, had to give his casting second vote to get the project through. CASM, Citizens Against Sand Mining, formed to demand a proper look at the project. They soon had 100 paid up members, 5000 signatures in disapproval of the scheme.

Peter Miller, the north coast project administrator, in an interview with ABC regional news described CASM as "a small minority waging a dirty and disgusting campaign of misinformation". They were asking questions, not making statements. One question was what was going to happen to the radioactive monazite if the company could not sell it? The company planned to return what was not sold to the mine site. All knew the experience of the neighbouring council, Byron Bay, which had had a very expensive cleanup of radioactive waste in the 1970s. Another company processing mineral sand in that area had offered their waste as land fill. Scores of homebuilders accepted the free dirt, even the local hospital spread it all over its grounds.

The other main question concerned the acid sulphate soil. How much of it was there? Since all they could get from the council was abuse instead of satisfactory answers, CASM challenged the project in the Land and Environment Court. "I am doing the job the council should have done in the first place," said Mister Justice Hemmings at one of the hearings. "The people have been given the mushroom treatment." In the end he gave the company the go-ahead but he applied ninety conditions on the mining that would not otherwise have applied, including the all-important appointment of a radiation safety officer.

The company began mining on 12 April 1990, but world prices crashed; it operated for two years, then pulled out. The mining did no harm whatever — it could not add to the awful damage to North Creek that the draining of the swamp had done. A proper assessment of the scheme by council, a reasonable explanation to the very worried people, would have saved ratepayers many thousands of dollars.

Another project that Ballina Shire Council is treating with autocratic obstinacy is that of a sewerage outlet at Lennox Head. The matter has already been before the Land and Environment Court, some civil action is probably pending as I write this. The north coast is in extreme trouble with its sewage. Little planning is being done, the problem has barely begun — it will proliferate with the increasing population. In a jocular reminder of the problem, the Clean Seas Coalition paraded the Big Poo through the streets of Ballina, an immense, fibreglass, yellow-brown turd mocking the town's Big Prawn.

Inefficient septic tanks are allowing dangerous levels of faecal bacteria into the rivers that are already suffering the same nutrient enrichment as western rivers. Coastal towns and cities are putting their sewage into the sea after various forms of treatment. They are so touchy about what they are doing that councils are even objecting to the term "sewerage outfall": "effluent outfall" sounds better. Coffs Harbour Shire Council is currently planning a new outfall to cater for its northern beaches. At the new Lennox Head plant all sewage will be treated by ultraviolet rays to kill bacteria before it passes into the ocean as unrecognisable sludge. Even if this treatment is as effective as they hope, it will do no more than solve the problem of recognisably polluted beaches.

These outfalls are pumping huge amounts of phosphorus into the sea which is not equipped to cope with it. There is no simple answer to the disposal of sewage, but pumping it out to sea not only wastes fertiliser, it wastes water. The first problem is to change attitudes to it, to regard it as a great natural resource carried by water that is becoming too rare to waste. Much can be used profitably for growing timber though this use is complicated by a tree's sudden reduction of its phosphorus requirements when it is about fifteen years old. Continued application at the same rate would kill the forest. But Memtec, a Sydney company, has devised a brilliant method of membrane filtering that can remove everything that there is in sewage in successive stages. Numbers of countries overseas are using their equipment. Australia has to catch up. So if this method were used, there would be no worry with the forests. The sewage would be put through extra filters to remove the nutrients, and pure water would irrigate the forest. In order to handle the unwanted nutrients, it is essential that scientists be set to work to perfect the drying of sewage so that it can replace superphosphate as a phosphoric fertiliser. The benefit would be enormous.

The still great hardwood timber resources of the north coast attract the attention of conservationists who want to shut up the whole area, of millers who want to continue logging, and of Pulp and Paper Manufacturers of Australia who want to build a pulp mill on the Clarence River. Discussion began in 1989. Meetings of up to 1000 people at Grafton unanimously opposed the mill, 12 700 signed a petition against it that was presented to Ian Causley, Minister for Natural Resources, who holds the Clarence seat.

Greenpeace declared that such a mill would pump 5 tonnes of organochlorides into the Clarence River each day. That organisation is notorious for exaggerating figures but a great trouble with such projects is that honest figures are not available. Neither government nor companies will release them, they prefer terms such as acceptable or safe levels. There is no safe level with organochlorides. They build up in aquatic food chains until what seem insignificant amounts become lethal.

Until the middle of 1993 the people of the Clarence valley believed that the petition had killed all idea of a pulp mill. Then the Northern Rivers Regional Development Board began a study to find out what timber was available. A pulp mill would destroy the Clarence River. There is not yet any technology that can prevent that happening. Jim Croft, president of BEACON, studied pulp mills in Canada that use the best possible methods. Each of them has turned a once beautiful river into a drain, not because of the poisons used in bleaching but because of the organic residue that pours out in fine suspension. It foams and coats bankside vegetation with a heavy choking scum, it turns the river water into thick brown refuse that only eels can live in. No seagrass can survive in the estuaries.

Pulp mills should not be given access to trees from natural forests. Australia grows some of the world's finest timber. It is vital that it be milled for its quality not its quantity. Selective logging need not be destructive. To deny the world of Australian timber is to degrade life, not enhance it. I do not wish to sit on a plastic chair at a plastic table, nor on a Radiata Pine chair at a Radiata Pine table. Plantation timber lacks the magnificence of age and slow growth.

Although so much has been written about the destruction of Australia's rainforests, the 2 million hectares that remain constitute a much bigger remnant than that of any other community of flora. That does not depreciate its value. Rainforest trees carry ancient qualities which are sometimes of extreme value to medicine, they have filtered millions of years of soil and saved rare products. The true value of rainforest is yet unknown. Most rainforest should be left undisturbed; unlike other Australian forests it can cope with itself. Rainforest reacts to destruction by growing different plants. That is not a disaster — even cyclones are necessary to a rainforest, they rejuvenate it just as a drought rejuvenates grassland. But what applies to a forest of 800 hectares does not apply to an 8-hectare remnant where disturbance can affect a whole forest and change it for ever.

After a cyclone it might take 800 years for a forest to return to its former state; even the removal of half a dozen trees growing close together in a small area might influence the forest for 200 years. So any milling of rainforest has to be carried out with great circumspection. Should there be a ban on the milling of rainforest trees? Of general milling certainly, but not of very selective milling. To leave magnificent timber to die and rot seems equivalent to filling in an opal mine before any of the discovered opal can be taken.

Dr Alison Specht of Resource Science and Management at the superb new university at Lismore carried out a study of Big Scrub remnants. Her work excites her and she projects her excitement with great effect. There are thirty-two remnants ranging in size from 4 hectares to 200. Some have been logged, some are threatened with weeds, some have cattle camping in them, but all are valuable. She suggests extending them, planting inside them where they have thinned, planting the edges as a buffer against weeds, and those on rivers and creeks could be connected by corridors planted along the banks.

The planting of banks that were once secured with plants is essential. The trees and shrubs were cleared, grazing cattle keep the banks bare. They blow away, wash away, slabs fall off. All creek and river banks that once carried forest will have to be replanted, extensions from them will have to be planted so that eventually 20 per cent of original forest is restored. That is not a cosmetic operation nor even a way of ensuring survival of species — it is the only way that productive farmland and water can be increased to meet the growing demand of the world, the only way that the north coast can remain a beautiful place to live. Bare paddocks cannot produce to capacity.

More vegetation will hold back runoff water. There will be a constant higher flow in streams instead of flash floods. Dan's Creek, a tributary of Cooper's Creek that flows into Wilsons River, drains a narrow valley 10 kilometres long and about thirty small creeks feed it. In 1888, a dry year, it maintained an average depth of 1.2 metres; in 1990, a wet year, the average depth was 15 centimetres. Most creeks are in similar condition. Wells at Brunswick Heads are going dry.

The north coast will certainly have to cope with sand mining again. It is a desirable occupation, and in the light of modern living it is a vital occupation. But, future mining will have to be restricted to the 7000-year-old beaches, those inland from the present coastal dunes. There does not seem to be any way that beaches or living dunes can be mined without damage. Topsoil can be removed and replaced with its shrubs and seeds to grow as before but the sand is not the same for them to grow in. There is always some clay in the sand which suspends during the process that separates the minerals. When returned to the trench behind the dredge with the spent sand, the clay settles from the slurry at an even rate, forming a hard pan that interferes with drainage. Few heath plants can survive waterlogging.

Bitou Bush is the worst result of the first sand miners and the companies cannot be blamed for it. In the mid-1950s the Soil Conservation Service recommended this South African exotic to stabilise bared sand. Like most things foreign, it found Australia a delight. The plant has overcome huge areas. In many places it has climbed over a series of dunes and is advancing into banksia scrub which it soon kills, nobody yet knows how.

Water Hyacinth, an escapee from garden ponds, filled in the Richmond River between 1916 and 1918. It is awesomely vigorous: one plant can cover an area of 600 square metres in a season, its seeds are viable for fifteen years. At its worst there were long stretches of the river where no water could be seen. Photographs of steamers show them apparently plying across a green paddock. In 1916 the Hyacinth Eradication Board set up. It sent out barges manned by fifteen men with rakes to pull it out of the water and unload it somewhere to be burnt when it died. The job was impossible. A lucky severe drought let enough salt water up the river to kill all the weed, a providential flood washed the mass of dead plants out to sea. Patches of it still grow. They are sprayed before they can spread.

Major Innes brought lantana to Port Macquarie as a pretty hedge. It now hedges thousands of kilometres of roads. Every gardener, every farmer, planted exotic reminders. Those who cut down all the trees and shrubs upon a creek bank were likely to plant Weeping Willows to replace them. Camphor Laurel, planted in gardens and avenues, did not spread out of control until fruit-eating pigeons, short of natural food, developed a taste for its seeds. Now it is a weed out of control.

Direct attack against established trees cannot work, there are too many. The best way of preventing any greater spread is by creating unattractive conditions. Camphor Laurel is replacing Swamp Mahogany as it dies after the draining of swamps; it takes over neglected areas such as abandoned banana plantations and will follow a road into a forest. It will not come away in healthy forests, it will not compete against a complex planting of natives. So more native forests, more swamp, more careful placing of logging tracks, means less Camphor Laurel.

The exotic vines that overwhelmed the Wingham Brush are overwhelming the vestiges of native growth on stream banks, and in many places they are accompanying lantana in a march from roadsides into forests. The much-publicised Bradley Method of clearing exotics from least affected areas towards the heaviest infestations does not work with these vines, they grow too quickly. Clearing has to begin in the worst areas by painstakingly cutting through every stem and painting the cut with pure Roundup.

Used with caution, Roundup is one of the great modern discoveries. But weeds suggest hormone and poisonous sprays, and sprays are being used with senseless abandon by farmers, State Rail, local councils, Northern Rivers Electricity, the Roads and Traffic Authority. That the work is useless is demonstrated by the growth of weeds along roadsides and railway lines that have been sprayed over and over. Sprays should be regarded as a dangerous tool to be used once or twice only, not as a casual remedy. Cane growers and banana growers, once regarded as immoderate and irresponsible in their use of sprays, now use them with increasing caution. For bananas a refined oil has replaced the dangerous fungicide used for years. The profitable market for organic produce is a fine incentive.

With increasing population, houses are now being built on and around former cattle dip sites, where the surrounding land is poisoned with arsenic and DDT. A committee established to examine cattle tick dip sites on the north coast published a report on their management in March 1992. It found 1607 sites, 1041 of them still in operation. Some of them are extraordinarily close together and a major difficulty for increased settlement. The topographic Wardell map south of Lismore covers an area of about 24 kilometres by 14 kilometres at a scale of 1:25,000. It names fifty-six dips. Cattle have to be dipped — apart from the fact that they cannot thrive without it, the law makes it compulsory — but this means these now awful areas are 2 to 3 kilometres apart.

Arsenic was used from the first dipping last century to 1955, DDT from 1955 to 1962. The Environmentally Hazardous Chemicals Act 1985 set the safe limit of DDT at 1 milligram per kilogram of soil, of arsenic at 20 milligrams per kilogram. In order to cut down on the difficult and expensive cleanup, in 1993 the Environment Protection Authority began a move to increase the safe limit of DDT to 50 milligrams per kilogram, of arsenic to 100 milligrams. The trouble is, that according to the committee's report, "the degree of exposure individuals could face from living in a house built on an old dip site is unknown". Anyway, most dip sites give readings way above any reckoning of safety, ranging up to 106 000 milligrams per kilogram of soil for DDT and 3000 milligrams for arsenic. There are forty-one houses now known to be built on twenty-four former dip sites. At present the only feasible way to deal with the contaminated soil is to dig it out and replace it with clean soil. But where can the contaminated soil be disposed of? No one wants it anywhere near them.

Part of the reason that I made this study of the north coast was to find out what effect Club Med will have if it is given approval to set up at Byron Bay. There is considerable objection. The town has an air of relaxed superiority about it. It looks good because buildings have been kept simple, there is no concrete high rise; it sounds good because the drumming and the sighing of the surf surges over it; it smells good with the sharp tang of incense cutting through the heavy perfume of melaleuca and acacia. The business houses are small and mostly owner-operated.

To find out the effect on local people, Elaine and I flew to Club Med in Moorea, one of the Polynesian islands of the Tahitian group for the minimum three-day stay. The scenery is truly magnificent, the mountains are stupendously steep and jagged, no postcard has exaggerated the beauty of the fringing lagoon. The accommodation huts in the Club's grounds are not obtrusive. Built of timber with shingled roofs, not ceilinged, they fit well among the coconut palms. ATTENTION AUX COCOS advises one notice and one does indeed have to be aware. During one day of exciting heavy wind and rain, a *maaramu*, a south wind, hundreds of coconuts came hurtling from heights up to 25 metres. We stayed inside and read, but the rooms are dark and the lighting atrocious. One is not meant to read at Club Med Moorea, there are instructors and provisions for many sports. Members come from America with racquets and cylinders of their favourite brand of balls to play competition tennis day after day. Others practised aerobics with a very tall, very good-looking, very athletic black African. There is swimming, diving, card-playing, volleyball, disco dancing until early morning.

The food in the main dining hall that can seat seven hundred (and often does) is astonishing. It caters for every taste: hot, cold, Polynesian, French, Japanese, general. One can select superb food or join the line at the hamburger bar. But after three days we began to feel secluded. Since all meals are paid for as a package with accommodation, it seems extravagant to go somewhere else to eat. The boundary of Club Med is a 3-metre mesh fence with the barbed top leaning out at a deterrent angle. The main gate and both possible beach entrances are guarded by *securité*. Club Med does not allow anybody through the grounds to the beach.

We left Club Med for the Moorea Village Hotel, simple, airy cottages with woven bamboo walls and pandanus roofs scattered about a lovely garden of the exotic plants that replaced Polynesian natives hundreds of years ago. There was no boundary fence. Our cottage had a kitchen, we hired a car and bought food from fishermen, roadside stalls and markets, taxing our French for the directions about how to eat it. For the first time we felt part of Moorea. Elaine said Club Med felt like a cross between a Butlins holiday camp and an internment camp. We expected to find resentment among the Polynesian people about this enclosure of foreigners in their country. To our astonishment that is not so at all. The Club has been operating since 1960 and the economy of the island now depends on it. It employs 200 local people, it buys a lot of fish from local fishermen, big quantities of fruit and vegetables from island farmers, pork from Tahitian growers. During its thirty-three years of operating it has closed down a couple of times for periods of two to three months. Moorea then came to a halt, even the taxis stopped running.

But the economy of Byron Bay will suffer, would have suffered (the decision is not yet made) by the advent of Club Med. Since the club would be setting up on land that is already in use as a holiday camp there would be little impact on the natural environment. The impact would be social, but that could be disastrous enough. Unlike Moorea, Byron Bay would probably lose money with the arrival of Club Med. It would make the town unattractive to the tourists and backpackers who now go there. The majority of backpackers are superior travellers, young, intelligent, adventurous, thoughtful. They come for four to six months and spend $45 a day, most of it with a number of shopkeepers. During their stay in Australia they spend almost $7000 in the towns that they visit. Japanese tourists spend about $2700 in an eight-day visit but the bulk of that is paid to overseas travel agents and Japanese duty-free shops. Members of Club Med buy airfares, meals and accommodation as a package. They would spend very little extra money in Byron Bay. They would be segregated, alien visitants.

Massive developments are likely to be thrust on any part of the north coast. With the present state of government they are likely to be welcomed without proper consideration of their effect. There have been outlandish blunders in recent times. In the 1980s Fruit Australia and Johnson's Farm Management cleared hundreds of hectares of private land near Glenreagh on the Orara River and planted it with fruit trees. The firms had no idea what they were doing, the nashi was the wrong variety for the temperature of that area, the land had been described by an inspector in the 1880s as "miserable poor country" and it had deteriorated since then. Until that work the Orara River had beautiful deep waterholes and many fish. There has been so much runoff from abandoned orchards that the river has filled with sand. The damage has been compounded by subdivision of unprofitable farms into rural residential blocks of 20 to 80 hectares. Clearing and new roads have caused wind and water erosion.

There is likely to be a particularly damaging housing estate built in an area known as North Ocean Shores, a mix of open wetland, rainforest and swamp that feeds Marshalls Creek, also known as the north arm of the Brunswick River. There is already a housing estate on the lower reaches of Marshalls Creek which includes a canal development, but the creek and the swamps to the north are still valuable fish-breeding waters and the forests are rich with wildlife.

There are moves for a big housing estate in swamp and forest at Iron Gates on the Evans River, for an urban village at Broken Head which would destroy the valuable remnant forest in the Broken Head Nature Reserve, for a community of 7500 people in the "reclaimed" swamps and mangroves at the Coraki Broadwater to the west of Tweed Heads, for communities on wetlands on the Nambucca and Hastings rivers where unauthorised and destructive work has already taken place; there are plans by predators for almost everywhere that one can go on the north coast. Part of the trouble is the senseless rivalry between towns that began a hundred years ago and continues. Population means power.

A size of 2 hectares has been suggested for the splitting up of rural land into concessional building blocks. It gives households the necessary room to dispose of their waste. The north coast cannot afford to put so much land out of production. An ideal solution would be for companies to buy groups of farms totalling about 1000 hectares, to plant 200 hectares of the area with native forest or revert it to swamp if it were once swamp, then to divide 60 hectares of the worst agricultural land into 500 building blocks, allowing each the old quarter-acre measure. The complex would handle its own garbage, its own sewage. The remaining 740 hectares would grow whatever was most suitable for the district. Such a scheme would be profitable, it would permit population growth and at the same time increase production. But only directors and managers with a deep knowledge of agriculture and practical economics could do it.

Most of the buildings in towns and cities show no aesthetic awareness. They are functional and ugly, architectural nightmares. In the 1840s Port Macquarie was described as "perhaps the most picturesque town that could be imagined". It is now singularly ugly, most of the old buildings are gone. There is a canal development which is now being extended. Such misuse of estuaries is as pernicious as drainage of bordering swamps. St Thomas Church of England, built in 1842–48, shows Port Macquarie as it was.

Grafton is an exceptional city, planned for beauty and maintained. An urban conservation area preserves the beginnings and the development of the heart of the city, and a long zigzag double-deck bridge (cars on top, trains below) joins the north and south halves across the Clarence. There are houses from 1850 and church buildings designed by J. Horbury Hunt, and the two main streets are wondrously wide, 2 chains by the old measure, now 40.24 metres, double the usual width. Even lesser streets are 27 metres wide. Despite its width, Jacarandas planted early this century arch across Pound Street. Since 1935 the magnificent avenue has been the subject of a yearly festival at flowering time. There can be no regret that the trees are exotic, they are part of Australia's white adolescence.

In 1989 the Department of Planning put out North Coast Design Guidelines, superb advice by an anonymous author. "Typical buildings ... would be light and airy ... covenants that require extensive use of brick or prohibit 'lighter' types of materials in house design e.g. galvanised iron and wood are not advised." The book provides many illustrations of how buildings can be sited so that they do not interfere with views up or down, it gives plans for imaginative blending of buildings with trees, shrubs and understorey.

Yet people are still building alarmingly pretentious houses on river banks, then clearing a couple of hundred metres of mangroves in front of them. There are many brick covenants on the north coast. If you ask councillors why they are ignoring expert advice they are likely to say that they "are trying to maintain standards — if we stopped to read all the bloody rubbish that is put out we would never get anything done". The big brick houses built in such areas do not look comfortable, they do not look welcoming, they do not blend into the landscape, they stand there blaring "I cost a lot of money".

Glen Murcutt, the architect, stipulates that houses should be unnoticeable. His remarkable buildings in wood, glass, galvanised iron and steel would be unacceptable to many north coast councils. There are two superb examples of his work up there: the Kempsey Museum which houses a display worthy of its setting and a house that he built for a family named Short on a farm on the Maria River near Crescent Head.

Architects come from all over the world to see the Kempsey Museum. In the last couple of years they have come from countries expected and unexpected, from Iceland, Japan, Hawaii, the United States, Peru, Alaska, Ecuador, Brazil, Holland, Germany, Russia, Finland, Norway and Sweden.

John Verge, the fashionable and excellent architect of the 1830s, retired to a farm on the Macleay River. Now Glen Murcutt has bought the Short farm, Victoria Valley, and he uses the house he designed as a holiday home. The roof is galvanised iron, walls and floor wooden, sun pours through skylights in the coved ceiling but it can be shut out with easily manipulated blinds if it becomes too obtrusive. Wide, heavy doors pivot or slide, double metal louvres control the entry of wind through the huge window openings. It is a long, wide house, about 27 metres by 11, yet if one looks at it from a distance a nearby clump of trees is the more noticeable. A former tack shed with heavy squared uprights that he turned into a flat has a front verandah that lifts up instead of turning down. It looks welcoming.

There has been immense interference with beach fronts, directly by flattening dunes and indirectly by building groynes and breakwaters.

There is sand where there was no sand, such as the beach at South Ballina which did not exist before the building of a breakwater in 1900; there is a similar beach at Fingal south of the Tweed. At Lennox Head the sea is 50 metres closer than it used to be. Sand dunes protected the village. Now the dunes are gone, together with a camping ground, a car park, two tennis courts, even the original Pacific Parade, the main street. Extreme seas coupled with extreme floods, such as those caused by Cyclone Pam in 1974, did millions of dollars of damage to houses and streets because the natural protection from the sea was gone.

Tourism is important to the north coast — it would not be exaggerating to say that it is vital to the economy. So for that reason alone, rivers and beaches must be kept attractive. No one will come to see canal developments on dead water. Agriculture is important and ought to be more important — there is great potential for offering unpolluted products, especially herbs, to European markets. But agriculture has too often been exploited by big money and little sense. Avocado plantings exceeded the possibility of any markets being able to handle the fruit, macadamias also were overplanted, then Hawaii took away the world market. Dairying suffered in the 1980s when butter consumption fell to one-third of consumption in the 1950s and overproduction in Europe meant loss of export markets. It is possible that people can be re-educated to eat healthy butter instead of unhealthy margarine. But there are still thriving pockets of dairies on good country, such as at Raleigh south of Coffs Harbour or at Dorrigo. Some of the modern dairies milking 200–300 cows do not graze them at all; when not in the milking shed they stand in concrete yards eating prepared rations. Despite the fact that the world's greatest cheese, Parmigiano-Reggiano, is produced from cows chained in stalls, it usually takes natural grazing to produce the finest flavours.

Beef cattle are particularly important and Casino even calls itself the Beef Capital. The standard of cattle has increased remarkably since the 1960s when a bounty encouraged widespread use of superphosphate. What damage it did and is still doing to the coastal rivers has not been estimated, but there is no doubt that it improved the cattle. Those of the 1990s dwarf the cattle of the 1960s.

Pockets of banana growers still make a good living. Coming into Coffs Harbour from the Coramba road there are spectacular steep hills of bananas, each palm stayed with three or four splayed props.

The Thursday Plantation at Ballina has made a great success of marketing tea-tree oil worldwide. Learning that Australia's production of 100 tonnes a year could not satisfy markets, Main Camp Station at Rappville, south of Casino, began the biggest tea-tree project of all — 15 000 000 trees on 2000 hectares. Will tea-tree oil go the way of avocados?

Maize can still be profitable, as are cut flowers and soybeans. The mixed soils and temperatures of the north coast offer so many opportunities. In an interesting move back to the lawlessness of the 1830s, marijuana has become a principal crop. The absurd laws that make it illegal make it especially profitable. Most of the sales are local, though some finds it way to the thriving market on the Queensland Gold Coast where the soil is too poor to grow it. It is a healthy criminality, with none of the ugly enforcement by syndicates that there is at Griffith where a grower who wants to pull out faces threats of a cruel death. On the north coast a grower might produce enough for a household and a little more to pay the rent or the rates, a cane farmer might grow a hundred plants for three years until there is enough to build a new house. First year cane makes an ideal cover crop, high enough to hide the marijuana but low enough to let in sunlight. Some go to a great deal of trouble to hide crops from helicopter searches by training the plants up trellises with beans or pegging them along the ground with pumpkins. Somebody stretched a net between two trees on the bank of the Brunswick River, covered it with plastic, piled soil on top and grew marijuana in it.

So there is the north coast. Its choice is an overpopulated desert or a populous and productive region of extreme beauty. The choice has to be made now — even tomorrow is getting late.

More a new planet than a new continent

When Europeans came to Australia, the soil had a mulch of thousands of years. The surface was so loose you could rake it through the fingers. No wheel had marked it, no leather heel, no cloven hoof. Digging sticks had prodded it, but no steel shovel had ever turned a full sod. Our big animals did not make trails. Hopping kangaroos moved in scattered company, not in damaging single file like sheep and cattle. The plentiful wombats each maintained several burrows so there were no well-used runs radiating from one centre as from a rabbit warren. Every grass-eating mammal had two sets of sharp teeth to make a clean bite. No other land had been treated so gently.

Much of the soil was deficient in phosphorus and some in trace elements like cobalt and copper, but plants and animals were accustomed to that, even appreciative. Nearly all soils, even the phosphorus-rich black soils, were short of nitrogen. Scattered legumes were so important most of them protected themselves with various poisons.

What did the country look like when the white settlers arrived? It looked superb. "New South Wales is a wildflower garden," wrote Watkin Tench, who came with the First Fleet. "The country is very romantic, beautifully formed by nature," wrote Elizabeth Marsden, wife of the Reverend Samuel, in 1794. "New South Wales has the appearance of a gentleman's park in England" wrote someone else. Henry Turnbull, out with Leichhardt in 1847, was first to describe the Springsure–Emerald country in Queensland:

> For upwards of 100 miles we passed through a most splendid, open country consisting of plains and downs — plains stretching out as far as the eye could reach on one side, and beautiful grassy slopes running down from a long and high range of mountains on the other. With only a tree to be seen every 500 or 600 yards, the whole face of the country was covered with the finest grasses and richest herbage, with wildflowers of every tint and colour.

Even the rather sour Daniel George Brock in the Barrier Range with Sturt in 1844, revelled in Australia's flowers. "As the day closed in," he wrote, "it was really delightful, the delight arising from the fragrant odours produced from the multitudes of flowers and herbs which strewed our path."

These descriptions of those who first saw Australia are important. I have found many, some still in manuscript, and nearly all show their admiration for the country. That attitude has been lost, clouded over by later literary assessments of men like Adam Lindsay Gordon, who was a bad observer, or Henry Lawson, who was a brilliant short story writer but who saw the land through the eyes of settlers on insufficient areas trying to make a go of it with money borrowed at 20 per cent.

The commonest remark of those practical white men who first saw Australia was: "You won't have to clear it to cultivate it." Australia was not a timbered land that had been cleared. "Everywhere we have an open woodland," wrote Charles Darwin when he came here in 1836. "Nowhere are there any dense forests like those of North America," explained *Chambers's Information for the People* in 1841.

One sees modern references to the clearing of the Hunter River country as though the first settlers put the axe into a huge cedar brush. John Howe and his party first walked along it in 1820. He reported: "We came thro as fine a country as imagination can form and on both sides of the river from upward of 40 miles [65 kilometres] (I may say) will at least average two miles [3 kilometres] wide of fine land fit for cultivation and equally so for grazing." He never mentioned the Cedar Brush. Certainly it was there — beautiful dense mixed growth with a huge Red Cedar every hectare or two. These brushes were a kilometre long by a kilometre deep, perhaps 2 kilometres long by 5 deep, never bigger, with many kilometres of open country between. The trees were no more than accent marks on open country.

Blaxland, Lawson and Wentworth could not have found their way over the Blue Mountains as soon as they did if the country had carried the present dense growth of tall eucalypts. One can no longer look across a valley to see where a scalable spur begins. Some of the heaviest forest of Victoria's Gippsland was open country at the time of settlement. When the rather rascally Lieutenant Jeffreys drove four-in-hand from Hobart to Launceston in 1814 before there was any proper road, he saw luxuriant pasture all the way and said there were few trees to impede his progress. Henry Hellyer named the Hampshire and Surrey Hills in north-west Tasmania after the grassy rolling downs he knew at home. They are now dense forest. Cook's Grassy Hill at Cooktown has become scrubby hill. If one goes into the densely timbered country east of Tenterfield in northern New South Wales and traces the long aqueducts cut by Chinese miners last century, one realises that, if the present growth had been there, the work would have been impossible. They could not have seen far enough ahead to peg their lines, roots would have made digging impractical. On the Palmer River in north Queensland, early gold wardens and geologists stressed the shortage of timber for mine props, boiler fires, even for camp cooking. The country now carries a thousand trees to the hectare. Present Australia grows many more trees than at the time of white settlement. Unfortunately the growth is disorganised. There are far too many trees where there ought to be few, such as the wide strip of ruined grassland north of Cobar in New South Wales, there are no trees where there ought to be many, such as the recharge areas of Victoria's Mallee.

What kept the country open? For thousands of years, Aborigines cultivated the soil with fire. The burnt land started with tree seedlings just as it does now, but the plentiful rats, possums, bandicoots and rat-kangaroos relished them and ate most of them. Aborigines burnt accidentally because they never bothered to put cooking fires out, they burnt as the incidental result of doing something else, rejuvenating a firestick for example, but most of all they burnt deliberately. One can scarcely exaggerate the extent of this management. In the central deserts they burnt every several years. The tribes walked over big areas and they burnt ahead of promised rain so the fires did not burn too much. The idea was not only to feed kangaroo and emus on the fresh green shoots they prefer, but to locate them. It was successful husbandry. The Desert Bandicoot and the Hare-wallaby disappeared when Aborigines went into mission stations and stopped burning. These small animals could neither move nor feed in deserts of prickly dead grass. Without fire to rejuvenate it, that country dies with startling suddenness. Dr Ken Johnson of the Northern Territory Conservation Commission found that the last Aborigines to come in from the Centre to a dreary mission station knew the Pig-footed Bandicoot up to the late 1950s. Thought to have become extinct last century, it did not die until the last fire was lit. In that country lightning strike does not happen often enough to support the desert life that depends on green plants.

In the higher rainfall areas husbandry was more intensive. Throughout the year tribes burnt successively along streams and around waterholes. Controlling the spread of the fires with bough beaters, they torched most of the land once a year, the most productive two or three times. The fires were mild grass fires, not the horrific forest fires of today. In the south-west corner of Western Australia several tribes met together and made a yearly ceremony of burning the perimeters of the Karri and Jarrah forests, the only big areas of eucalypts in Australia before white settlement. They appointed lighters with firesticks, they appointed bearers with green boughs. Aborigines did not like forest. It was uncomfortable to walk through naked. It fed less game. These great forests extended very quickly after white settlement disrupted Aboriginal life, then European axemen moved in.

Some tribes cultivated plants by fire. The Gumatj people on the western shores of the Gulf of Carpentaria still control the fruiting of cycads. These plants bear pineapple-shaped cones that are a staple food even though they are a deadly poison before washing and fermenting. If left alone the cycads would produce a scant crop lasting a few weeks only. Fire encourages fruiting so a succession of planned fires produces an abundant crop over several months.

The spongy original soil took in water quickly and released it slowly. Streams in all the better rainfall areas kept a more certain flow. Watercourses like the Mooki River in north-west New South Wales that explorers described as "noble streams" now look like dry gullies for months or years at a time.

The original grasses were mostly deep-rooted perennials that grew in distinct clumps. They had to be deep-rooted because of the nature of the topsoil. Oat Grass, *Themeda australis*, was the principal grass of both coast and inland. It stretched from southern Tasmania to Western Australia's Kimberley. It is so beautiful that members of the Society for Growing Australian Plants now cultivate it in Canberra gardens. Some grows on our own farm, but stock do not like it much any more. Ample phosphorus and nitrogen have changed it from a medium-height, soft, fine-leaved grass to one that is tall, thick-stemmed and coarse.

Among the grasses grew tap-rooted herbs and many other plants. But always there were bare spaces. Birds and small mammals had plenty of room to walk about while they fed on fruit and shoots or stretched upwards for the oat-like and millet-like seeds of the grasses. This lovely pasture lasted about six years in most districts. The method of stock management hastened the destruction. Stockmen mustered their cattle each late afternoon and camped them down on water, shepherds yarded their sheep and drove them out again each early morning. The ground powdered under the cutting hooves, then hardened when it rained. The plants had never had to push their roots through hard ground, they had never had their leaves bruised by cloven hooves, they had never had whole bunches of leaves torn off between a set of bottom teeth and a top jaw pad. They died. Bare ground ringed out from camps and yards. When the stock had to travel too far to feed, yards and camps were shifted. More ground was bared.

Ground never stays bare for long. Inferior Australian grasses, the vicious corkscrew-seeded species of *Aristida* and *Stipa* which had grown sparsely on rocky hillsides, found the new conditions ideal, so did imported weeds with thousands of years' experience of hard soil. Settlers first took their stock to the lovely Hunter River in 1821. By 1826 they had eaten the country bare. In 1859 botanists inspected it for the New South Wales government. On farm after farm they found no Australian plants. All that grew was imported weeds.

What manner of men caused this destruction? They were not greedy and ignorant, many of them had a background of hundreds of years of good farming. They could usually estimate pasture, its stocking rate and recovery time, but it was beyond human achievement to assess this land correctly. It was more a new planet than a new continent. It developed to an advanced state without the modification of humankind: no intelligence impinged on its isolation until Aborigines began to work it about 120 thousand years ago. I believe that if a team of modern scientists with 200 years of new knowledge were presented with Australia as it was in 1788, they would make almost as many mistakes. It was even more difficult in the low rainfall country. No one knew what the rainfall was and areas of 100 millimetres a year looked exactly the same as areas of 200 millimetres. The difference was in the recovery period after grazing.

Much of the later destruction, that from the mid–1830s to the 1850s, was due to bad land laws. Graziers could not spell any parts of their runs. If a commissioner or land inspector found an area without stock, he immediately declared it unoccupied and allotted it to someone else. A few of them were well paid to keep their eyes open.

The astonishingly quick change to our original pastures is little known, mostly because the evidence for it is still in manuscript and hard to find. The great area of speargrass in Queensland is usually regarded as something that was always there. In her *Cry for the Dead*, Judith Wright told how that worthless species came down off stony hills and overran what had been the rich Dawson River valley as soon as the first good grasses were trampled out. Most landholders on the Liverpool Plains regard the Plains Grass, *Stipa aristiglumis*, that grows there so prolifically as the original pasture. Cunningham, a fine botanist, listed the grasses he found there in 1825 and did not mention it.

Despite the present abnormal number of trees, it is necessary to plant millions more as highly efficient pumps to lower water tables lifted to disastrous levels by ignorant irrigation. Provided shrubs, understorey plants and grasses are planted among them to encourage insects, spiders, birds, small mammals and the vast host of minute life, these new forests will eventually help to restore southern Australia. It is not trees which are missing, it is the shrubs, deep-rooted perennial grasses and the many flowers and animals that go with them. Huge areas which look superbly treed from the air will not stand inspection on the ground. There is little native life there. A second-rate European pasture grows among ageing Australian trees.

For many years the principal fodder plants in southern Australia have been Barley Grass, *Hordeum leporinum*, and Burr Medic, *Medicago polymorpha* var. *vulgaris,* a plant English farmers referred to slightingly as Black Clover. Traditional old Australian graziers look out across their paddocks of cattle fattening on these two plants and say, "You can't beat natural pasture". In old reports and old diaries you can watch Burr Medic spreading. Men noted it because the plentiful coiled hairy burrs spoilt the wool clip. By 1859 it was on the Macquarie, the Lachlan, the Castlereagh, it was spreading in the Hunter and already edging on to the Liverpool Plains. By the 1890s it appeared on the great blacksoil plains between Narrabri and Moree, and there it had an extraordinary effect. That country was very rich in phosphorus, low in nitrogen. Several species of rats were common, Flock Bronzewing Pigeons fed there in huge numbers. They nested on bare ground between the grass clumps, one bird sitting a few centimetres from the next. Old-timers tell of sheep's legs yellow with broken yolk. The rats and the pigeons were trampled out in a few years. By early this century, Burr Medic had filled in all the spaces. It supplied nitrogen in abundance. Most of the original grasses lived on in that country. Self-mulching soil cannot be trampled permanently hard like the red soil. But the dominance of the grasses changed and many more annuals grew. The whole pasture thickened. It now feeds many more stock than it did a hundred years ago. But the rats and the pigeons could never come back. There is nowhere now for them to walk about while they feed, nowhere for the pigeons to nest.

Kangaroos had a difficult time of it for the first seventy years. They were not originally in great numbers. Oxley once reported hundreds, so did Gilbert in Western Australia. Most other explorers in country where one would expect to find many either reported few or none at all. Leichhardt in his journey north in 1844 expected to live off the country. He soon noted in his diary: "It has now become painfully evident to me that I had become too sanguine in my calculations." And Leichhardt was willing to eat almost anything alive.

Kangaroos at the time of white settlement were well controlled by Aborigines and dingoes. Everything in Australia was in a state of balance. That does not mean a static mixture of ingredients as in a baked cake, it means a state of great tension where each species is controlled by its association with other species. The first hungry settlers quickly changed the orbit of the kangaroo. They hunted them first for their flesh. Even Governor Arthur Phillip served kangaroo on 4 June 1788 when he celebrated the King's birthday by inviting the officers to dine with him. The surgeon of *Sirius*, George Bouchier Worgan, recorded the meal in his tiny handwriting:

> About 2 o'clock we sat down to a very good Entertainment, considering how far we are from Leaden-Hall Market, it consisted [sic] of Mutton, Pork, Ducks, Fowls, Kanguroo, Sallads, Pies and preserved Fruits. The potables consisted of Port, Lisbon, Madeira, Teneriffe and good old English Porter, these went merrily round in Bumpers.

It did not take long to realise how good kangaroo skin was, too, both as rugs and as leather. Skin-getters went to work. They exported pelts by the thousand. When the young Charles Darwin went to Bathurst eager for a kangaroo hunt, no kangaroo could be found for him to hunt. "It may be long before these animals are altogether exterminated," he wrote, "but their doom is fixed." Five years later, in 1841, John Gould, the great ornithologist, wrote:

> Short-sighted indeed are the Anglo-Australians. Let me urge them to bestir themselves, ere it is too late, to establish laws for the preservation of the large Kangaroos ... without some such protection the remnant that is left will soon disappear.

Australians did not bestir themselves, the kangaroos did. New permanent waters put down for sheep and cattle allowed them to extend their restricted natural range. Everywhere sheep cropped the grass to the after-fire length the Aborigines had so long prepared for them on small areas. And suddenly all predation was lifted. Skin-getters pulled out because they were getting too few skins. Landholders stopped shooting because the few kangaroos left no longer damaged crops. Most of the Aborigines were dead or growing fat and sick on white flour and white sugar. And by 1860 the hunting and poisoning of Dingoes had been so successful that they were exterminated on most settled land. Kangaroos found themselves in a wonderful new world with all they could eat and drink anywhere they liked to go, and nothing to eat them. They took full advantage of it. All over southern Australia, kangaroos increased extraordinarily. Alarmed landholders retaliated against them in every colony. They built big winged yards and men rode in from neighbouring stations with dogs, guns and whips to drive them into them. The old hunting term, "battue", came into common usage. Newspapers recorded the yarding of 3000 on one day at Geelong, 14 thousand in two years at *Outalpa* in South Australia, 16 thousand on *Oulnina*, the adjoining station, over 61 thousand in twenty months at *Gordon Downs* in Queensland, 80 thousand in a few weeks at *Trinkey* station, near Gunnedah. "It was a systematic attempt at the extermination of the tripod," wrote one journalist.

The 1860s are important in the environmental history of modern Australia. Much change took place then. The "very nice-looking young ladies" who startled Darwin at a Macarthur dinner party by exclaiming: "Oh, we are Australians and know nothing of England" had, by that time, colonial sons old enough to want their own blocks of land. There was a clamour for closer settlement. And many of the older men had made money enough to think of the hunting and shooting they had left behind in England. There had been many attempts to bring in game animals: rabbits from the very beginning, hares, deer, foxes for hunting. The first importations of foxes were all dog foxes and so valuable by the time they got to Australia that the hunt masters put collars around their necks so they could snatch them out of the jaws of the hounds and run them again another day. If a master was a bit slow at the end the loss to the club was enormous. So hunt clubs brought in vixens for breeding, so did private individuals. Then bodies called Acclimitisation Societies went into business. Australia sent magpies to England, England sent nightingales to Australia. And they sent starlings, sparrows, blackbirds, more hares and deer, trout, skylarks, creatures that thrived and did not thrive, the harmless and the harmful. Every ship's cook had a menagerie in his galley.

Rabbits brought to New South Wales and South Australia began to breed up after a long period of being held in check by predators. And rabbits in Victoria, mostly descendants of those nurtured so carefully by Thomas Austin of Barwon Park and the friends he gave pairs to, bred up with enormous success. A market began. Four and a half thousand rabbits a week were auctioned in Geelong, Ballarat and Melbourne. The attitude to rabbits changed completely. Landowners who wanted them for sport now cursed them as pests. Workmen saw them as the poor man's friend. "There goes many a poor man's dinner" someone would say as he let a pair go on new territory. The rabbit replaced gold. It became a symbol of independence. The farmhand could throw in his job and go rabbiting. That attitude persisted into recent years.

A long drought in the 1870s beggared everything: rabbits, kangaroos, native cats, rat-kangaroos, farmers, graziers, the soil itself. Rivers dried up, sheep and cattle churned the soil into dust, then died in their hundreds of thousands. When it rained and kept on raining trees and shrubs germinated in the bare ground just as they do after fire. There was nothing left to eat them. White Cypress Pine scrub sprang up down the Lachlan in New South Wales, Bimble Box on every little bare hill in the central west, Coolibah on the Warrego in Queensland, River Red Gum along the Murray and Murrumbidgee. Rainforest leapt out of its confines in narrow gullies on the north coast of New South Wales and climbed the hills. The Karri and Jarrah forests in Western Australia ringed out around the axemen. Brigalow clumps grew into Brigalow scrubs in Queensland, single wattles into tangled masses. The great Pilliga Forest in northern New South Wales leapt out of the ground and drove out men, sheep and cattle.

The growth destroyed many landholders. It was priceless for Australia. It came up so quickly it was cast like a protective net over plants, birds, mammals, insects that would have disappeared forever in the disasters of the next seventy years.

In the 1880s and 1890s rabbits spread, and were spread over the whole of southern Australia. If a trapper caught a pregnant doe, he let her go again. Overlanders walking out of South Australia across the Nullarbor carried billies full of furred kittens to be set down at a good waterhole. Crates of rabbits, fifteen to a crate, were consigned by rail to Queensland for release on the Darling Downs. In the next drought of the 1890s millions of rabbits in the Riverina swarmed and ran in search of water. They ate everything moist in their path, even Kurrajong trees up to 17 centimetres in diameter. In the north of South Australia they climbed trees like possums, ate the leaves and stripped the bark. In the Centre they turned Mulga scrub into barren plains and they have kept them barren ever since.

Pastures over huge areas degenerated to plants that rabbits would not eat. Sand and red dust lifted in storms that blacked out towns like Broken Hill for a couple of days at a time. Mothers hung wet sheets over their babies' cots to filter the air and they shovelled dust out of their houses by the bucketful. As an answer to rabbits, horse-drawn poison-carts trailed out blobs of a pasty mixture of pollard, water, molasses and phosphorus. In forty years of work they laid about 150 million kilometres of poison over southern Australia. Grain-eaters and flesh-eaters died in thousands. The rabbit population thrived. The poison killed enough to keep the breeding stock healthy.

The disruption to the thousands of years of undisturbed soil intensified with the development of farming. Single furrow ploughs drawn first by bullocks, then by horses, took the place of hoes. It is startling for a modern farmer to read of the simple working of new ground 150 years ago. A paddock could be ploughed with a single-furrow mouldboard, then left for weeks or months until sowing time when the weathered clods were broken down by a tined rake, the seed broadcast by hand and covered with harrows. The ploughing killed the native plants and there were no imported weeds to grow. But most early crops were for private or very local consumption.

Railways prodding out from capital cities into farmland permitted commercial grain growing. Wheat, oats, malting barley supplemented wool and beef and increased Australia's income. The stumpjump disc plough invented in Victoria allowed a disc to ride harmlessly over an underground obstacle instead of cracking its castiron arm. William Farrer, a plant-breeding genius, worked for twelve years without pay and without recognition to modify European wheats to suit Australia's reversed seasons, then to breed into the most promising a resistance to the fungus disease rust which turned almost the entire crop of 1889 into red mush. The variety Federation was his great success. It established the modern wheat industry. Hugh Victor McKay built a stripper harvester to reap, winnow and bag the wondrous new yields in one operation. Australia led the world in the development of modern graingrowing machinery, and farmers began work with it on an astonishing scale.

Mack and Austin at Narromine had seven four-furrow disc ploughs working one behind the other, each plough pulled by a five-horse team. On George Henry Greene's *Iandra* at Grenfell, sixty-one share farmers grew over 8000 hectares in 1909–10. About 700 men took off that massive harvest. Modern farmers efficiently replaced men with machinery and are no longer big employers of labour. So they lost their influence with politicians and their importance.

In the great Riverina of New South Wales, irrigators spread water with equal enthusiasm. For a time deserts bloomed even more prolifically, than anyone had dreamed of. But water was used too lavishly over too great an area. The underground aquifers and channels which drained dissolved salts into rivers at low summer level could not cope with a great new constant influx of excess water. They were equipped to handle only the little which bypassed plants in a rainfall of 350 millimetres. Water tables rose, bringing salt to the surface. A film made in the late 1920s shows soldier settlers of World War I stacking furniture and fowls on to horsedrawn drays. Then the camera pans over orchards of dead fruit trees in a bed of white salt.

By the end of the 1940s, southern Australia was generally a bleak place. World War II had interfered with work on farms and stations. Rabbits were out of hand again. One saw few wildflowers, few birds, few mammals. The land itself looked sick. I had lived all my life on a farm and had never seen a kangaroo on my father's farm or on anyone else's within 50 kilometres. I had seldom seen a kangaroo.

Foresters stationed at Baradine in the great Pilliga Forest in the 1940s had never seen a White Cypress Pine seedling growing naturally. For sixty years, over more than half Australia, rabbits — even greater lovers of tree seedlings than rat-kangaroos — had eaten almost all that germinated. Then came the miracle of myxomatosis and a ten-year cycle of good seasons. With nothing to eat them, pines in the Pilliga came up 50 thousand to the hectare and all tried to grow. Wildflowers we had never seen blossomed in profusion.

The Burramys, a possum known only as a fossil, seemed to crawl out of the rocks. The New Holland Mouse turned up again, and the Pilliga Mouse (my son, Mitchell, caught the female that established the species), the Parma Wallaby, the Dibbler, other possums, bats, snakes, lizards. In the early mornings one came on koalas sitting on roadways licking up calcium-loaded sand where no koalas had been seen for fifty years. Another protective net, a low one of understorey, was thrown into our forests. Australian soil had come to life again. And the new forests, like arks, packed in all they could hold, even animals like the Red and Grey Kangaroos unaccustomed to living in forests.

Farming extended. Gabo, a new variety of wheat bred by Dr W.L. Waterhouse, another plant-breeding genius, opened up the black soils which were too rich for the old varieties. Farming had been confined to red soils. On black soils, wheats like Ford grew 1.8 metres high, began to head, then hayed off, fell over and yielded nothing. Every characteristic of Gabo fitted it for black soil. It was short in the stem so that summer storms did not flatten it, it was the most rust-resistant variety ever developed, it was high yielding, quick growing so that it could be sown into late August if rain was delayed, it thrashed easily, it was high in protein yet good to mill. These last attributes had been regarded as opposites. The kerosine-engined steel-wheeled tractors that had replaced horses were now out of date. New rubber-tyred diesel-engined tractors hauling new wide machinery and fitted with lights for twenty-four hours a day working moved on to hundreds of thousands of hectares of new country. It grew the best wheat ever seen. A series of good seasons added to the excitement.

The big kangaroos made immoderate use of the seasons. By 1958 portable freezers dotted the country and shooters drove in with rifles for the kill. Kangaroos are worth killing. Their skins make superb leather — the finest of all for shoes — their flesh tastes good and is low in fat, an excellent food for sedentary modern society.

Rabbits built up again, especially in the Centre. A big rabbit-dealer out on the Birdsville track told me in 1966, "Myxo's the worst thing ever happened to this country. Cost me God knows how many thousands." Then he rubbed his hands together. "But the country's just beginning to come good again."

Australia never comes good for many years at a time. Droughts are an essential part of the cycle. But it was never intended that soil should be bare during every drought. Kangaroos stop breeding on dry feed. When waterholes dried up, they died. The soil rested with a good dry cover and rebuilt its chemicals. Under sheep and cattle the western lands of New South Wales were laid bare in the 1870s, they recovered, they were bared in the 1890s, recovered, bared in the 1900s (you could flog a flea from Narrabri to Walgett across the plains), they recovered, bared in 1964–66, recovered.

Then there were a number of good years. By 1979 kangaroos had increased to numbers unhealthy for kangaroos, let alone for crops, pastures, sheep and cattle. A new harvest began, better controlled this time, though many shooters still shot for skins only, a waste of thousands of tonnes of good flesh. During this buildup, kangaroos made unusual movements. Eastern Grey Kangaroos moved farther west than had ever been known, Red Kangaroos moved farther east. And both Reds and Greys took up permanent residence in forests.

By 1981 the country was in drought again, probably the hardest, longest, most widespread ever experienced. Huge areas of western New South Wales were laid bare. Even trees and shrubs died. My information about dust storms had come from books. Now I experienced them, as did everyone else living from Melbourne to southern Queensland. The country recovered again as best it could. How many more times can it recover? As I write this it is trying to recover yet again from another long drought of 1991–92. The tortured ground throws up spectacular covers like the useless native Poached Eggs, or introduced Wild Sage. Mulga comes up too thickly on fortunate flats that collect some runoff water where it has never grown before and which ought to be growing grasses, and fails to grow where it always grew unless the area is fenced off from stock and rabbits. Several species of saltbush, once a dominant pasture, are gone from huge areas but they make a slow comeback in experimental fenced-off plots, and among them a few species of grasses make a miraculous resurrection.

So it is not too late to save that land, but it will require changes to laws, to attitudes, to the lives of some landholders; it will also require unaccustomed courage and good sense on the part of governments and some generosity to displaced graziers. After all, bad government is partly to blame for the state of the West where many graziers are failing anyway. After World War II it settled soldiers on cut-up stations and fixed a compulsory stocking rate calculated on carrying capacity over so-called normal years. It excluded droughts as though they were an occasional calamity instead of an irregular but certain necessity.

When in 1991 the great Darling River painted itself a blatant poisonous green from Mungindi to Wilcannia, it was protesting about our misuse of phosphorus. In a country as deficient in phosphorus as Australia, it is absurd that sewage and abattoir run-off, both containing particularly valuable nutrients, should be regarded as waste and discharged into rivers and oceans which cannot handle them. The metropolitan farm at Werribee out of Melbourne demonstrates the quality of pasture that sewage can grow. It would be even more valuable if towns used it to irrigate planted tree lots for pulping and for milling. There is enough to produce significant forests and ease the pressure on our natural forests. Woodchipping is a great industry, provided trees are grown for it as a commercial crop to be harvested like any other crop. It is a misuse of native forests.

At the time of European settlement our rivers were extraordinarily well equipped to handle rich silt. No precious nutrients flooded out to sea, but no single organism could dominate their richness and flourish out of control wherever they were deposited. An exceedingly complex world of plants, viruses, bacteria, aquatic animals of all sizes and waterbirds kept control through drought and flood. The spongy soil which controlled runoff also held back much moisture during long, dry spells. It was normal and necessary for streams to dry up to chains of waterholes every several years so that the underground salt streams could drain. The many dams and weirs have interfered with this drainage in places not yet subjected to too lavish irrigation and put even more country at risk of salting. Sheep, cattle and farming have destroyed reedbeds and removed the natural filters.

The country has to be reassessed, reorganised. In some places, such as the far west of New South Wales, consideration for the country has to be the main concern. At one-quarter of the present stocking rate it could recover in thirty to forty years provided rabbits, goats and wild pigs were eradicated or at least held in check. During the depression in wool prices of the late 1980s, some landholders began to round up and market up to 12 thousand wild goats a year. Rather dangerously, they now regard them as a handy resource. Native plants cannot tolerate the intemperate browsing and grazing of goats. Farming kangaroos and emus instead of sheep and cattle would relieve the far west of New South Wales, though kangaroos can be almost as destructive as sheep if confined in too great numbers.

Because of the damage to so much of Australia, our forests, our parks and nature reserves are now places of inordinate value. They were created by accident, the growth of the 1880s, they were saved by the almost accidental spread of myxomatosis in the 1950s. During the last few thousand years those two periods, 110 years ago and forty years ago, have been the only times when it has been possible for trees to come away as forests in most of Australia.

Life in the forests has intensified markedly over the last forty years. They are now exaggerated communities of plants, so the insects that feed on plants are in exaggerated numbers, as are the animals that feed on the insects. So animals that feed on insect eaters have increased, as well as those that eat nectar, fruit and seed. It is an accelerated and unnatural world, and it is about forty years old. Our forests do not display the past as it was, they have concentrated it. They are fragile, extraordinarily vulnerable, and they are being managed by accident.

At the time of white settlement no part of Australia was accustomed to looking after itself, not even the rainforests. Each year we learn more of the intensity of Aboriginal management. Our forests and parks require management according to their main function. If a park is to feed Yellow-Footed Rock Wallabies it needs successive burning several times a year, if it is to display some species of acacias then it needs burning every ten years or so, if a forest is to grow White Cypress Pine timber, any fire is damaging. Some parks should be enclosed with substantial double electric fences and dingoes introduced as a natural method of keeping kangaroos in check. This would also save the dingo, which is fast losing its identity in crosses with domestic dogs.

The management of the whole of Australia has to change, especially the methods of irrigation. Australian soils welcomed floods to dissolve salts and lodge them in aquifers to drain during dry years. It is impossible for them to handle regular flood irrigation. So water has to be rationed to crops in the exact quantity that they can use. It will be initially expensive but it will ultimately extend irrigation without salting any more country.

What is happening now in Australia and in the world is serious, but it is not terminal, it is a minor disorder. Everything that we are doing wrongly is correctable by us, they are human errors, not almighty ungovernable forces. Aborigines made this a very beautiful country. But consider what it must have looked like 120 thousand years ago when they began to modify it to suit themselves with blackened tongues licking into rainforest and into the widespread dense casuarina. Scattered eucalypts came up like weeds on the newly scarred land and grasses that no one knew were there began to grow among them.

The scars that we have made are the result of our attempts to mould the land to our own needs, and they look so terrible because they are so recent. They will look ugly for another hundred years or so. But we are learning, in some cases we have learnt. We, too, can build a beautiful Australia, but to do so depends on our maintaining a passionate interest in the environment.